363.7392 Breuer
B846a Air in danger
1980

Glendale College
Library

Air in danger

DISCARD

Air in danger
Ecological perspectives of the atmosphere

GEORG BREUER

TRANSLATED BY PETER FABIAN

CAMBRIDGE UNIVERSITY PRESS
CAMBRIDGE
LONDON NEW YORK NEW ROCHELLE
MELBOURNE SYDNEY

Published by the Press Syndicate of the University of Cambridge
The Pitt Building, Trumpington Street, Cambridge CB2 1RP
32 East 57th Street, New York, NY 10022, USA
296 Beaconsfield Parade, Middle Park, Melbourne 3206, Australia

© 1978 Deutsche Verlags-Anstalt GmbH, Stuttgart

Revised English edition © Cambridge University Press 1980

First published in German as *Geht uns die Luft aus?*
by Deutsche Verlags-Anstalt GmbH, Stuttgart 1978
English edition first published by Cambridge University Press 1980

Printed in Malta by Interprint Limited

Library of Congress cataloguing in publication data
Air in danger.
Translation of *Geht uns die Luft aus?*
Includes bibliographical references.
1. Air – Pollution. 2. Atmosphere. 3. Ecology.
I. Title.
TD883.B7313 363 79-18820
ISBN 0 521 22417 9 hard covers
ISBN 0 521 29483 5 paperback

363.7392
B846a
1980

5/81

Contents

Is our atmosphere endangered? vii

1 **Oxygen and carbon dioxide – wherein lies the danger?** 1
Are forests oxygen producers? 1
Forest destruction and carbon dioxide increase 7
Erosion and oxygen loss 13
Are algae oxygen producers? 17
Is atmospheric oxygen endangered? 20

2 **The history of the atmosphere** 25
General dimensions 25
What the atmosphere reveals about the Earth's origin 27
The ancestors of our atmosphere 28
'Air' without oxygen 31
Radiation 'currencies' and atmospheric 'customs duty' 34
Life by chance? 38
From fossil food to photosynthesis 41
How old is life? 44
Team work 48
Banded iron formations 50
A new environment 55
The origin of biological cycles 58
Is the long-term cycle balanced? 61

3 **Carbon dioxide increase – how great is the risk?** 67
Atmospheric carbon dioxide and fuel consumption 67
Eighty tonnes of labels 72
Is world biomass growing or shrinking? 76
What can be learnt from tree-rings? 81
The search for the missing sink 85
Forecast with many unknowns 89
Danger for the climate 96

4 **Nitrogen and ozone – unexpected relationships** 105
The origin of nitrogen self-supporters 105
Common ancestor or natural gene transfer? 108
The nitrogen cycle 111
Intentional and unintentional nitrogen fertilisation 115
Almost a noble gas 119
Spray cans, fertiliser, supersonic aircraft 125
Starvation or skin cancer? 130

5 **Between the devil and the deep blue sea** 134
If microorganisms were human beings 134
After the next ice age 138
A third alternative? 144
Is the carbon dioxide issue being overemphasised? 148
Always just 'one horse, one rabbit'? 151
A strategy to buy time 156
Reforestation – a policy with many benefits 159
After us the deluge? 166

Tables 171

Notes 174

Glossary of technical terms 184

Index 187

Is our atmosphere endangered?

The air we breathe is a product of, rather than a precondition for, life on Earth. In its present composition, it mostly originates from metabolic processes of living organisms rather than as a mere consequence of cosmological or geological history. Thus our existence is bound to a very special environment which is unique in the whole of the solar system.

The main components of the atmosphere are nitrogen (78 per cent by volume), oxygen (21 per cent), and the noble gas argon (slightly less than 1 per cent). In addition there is about one thirtieth of a per cent of carbon dioxide (CO_2), traces (less than one thousandth of a percent) of other gases such as methane, hydrogen, oxides of nitrogen, and other noble gases, and varying amounts of water vapour, which is usually considered as an admixture rather than a constituent[1] (see also table 1 on page 171).

Apart from water vapour, i.e. the varying humidity, the composition of the air is remarkably constant. Up to about 100 kilometres altitude the percentages of the main components do not show any variation with geographic latitude or season. It is only carbon dioxide which, apart from man-made effects, shows natural diurnal and seasonal variations that are caused by vegetation.

The stability of the composition of the atmosphere is all the more remarkable since, in a chemical sense, our atmosphere is not stable at all. According to scientists it is a mixture of gases which, while water is present, are not in thermodynamic equilibrium. Oxygen would, by oxidising minerals contained in the Earth's crust, in particular iron, carbon, and sulphur, be consumed in the course of some million years if it were not continuously replenished by green plants. If, on the other hand, the existence of oxygen is taken as a fact, chemical reactions with atmospheric nitrogen are likely to occur that would gradually convert all the waters on Earth to nitrate solutions. Thanks to certain species of bacteria that decompose nitrates and recycle the nitrogen back into the atmosphere, this does not happen.

The atmospheric constituents, permanently involved in a large number of geological and biological processes, are replenished through complex cycles with living beings playing an important role. Thus the composition of the atmosphere can remain constant only as long as all these cycles, with differing time-scales, remain in a state of equilibrium. It is obvious that such a complicated system, with only a limited ability to self-regulate, may be vulnerable to natural or man-made disturbances.

This book describes how our atmosphere, and the processes regulating its composition, have developed throughout thousands of millions of years, and how man has been interfering with these processes. It analyses how and to what extent this natural equilibrium might be disturbed by human interference and what might be the consequences for life on Earth.

It was concern about atmospheric oxygen that gave me the first impetus to deal with this complex problem. During the past few years, distinguished scientists have repeatedly pointed out that the global oxygen consumption, through various technical processes, is growing at an alarming rate, whereas the most important oxygen producers, forests and marine algae, are being destroyed through deforestation and oceanic pollution at a rate that is no less frightening. These opposing trends must lead one to believe that, in the long run, disaster must result. A thorough study of the available scientific material however yielded surprising information at least for the present. Contrary to what one would expect there is no reason at all for such apprehension. Oxygen is one of the natural resources that is practically unlimited, at least as far as time-scales of hundreds or thousands of years are concerned. The total mass of atmospheric oxygen of about 1200 billion (i.e. 1200 million million) tonnes is so enormous that, for such time-spans, no human activity whatsoever could cause any serious change. Nevertheless, this should not be used as a reason to stop worrying. The opposing trends mentioned above do not only affect oxygen but carbon dioxide as well, only 2.6 billion tonnes of which are present in the atmosphere. The same effect that causes only a slight reduction of oxygen, not measurable with present-day analytic techniques, causes a noticeable increase in the considerably smaller reservoir of carbon dioxide. Between 1957 and 1977, atmospheric carbon dioxide increased by 5 per cent, and a doubling or even trebling is considered to be possible in the course of the next fifty to seventy years. Most climatologists are of opinion that global warming and hence a shifting of climatic zones and precipitation areas would be the likely consequence. This might require difficult and costly adaption processes and possibly have a disastrous impact on agricultural conditions in many areas of the Third World.

It is certainly necessary to cope with acute problems of air pollution by noxious gases, dust and radioactivity; but these are not the subject of this book. However in dealing with these problems one should not neglect the global increase in carbon dioxide, which in the long run is likely to be an even bigger problem. Once the problem becomes acute it will be too late to start to try and solve it because changes in the pattern of energy production and use, probably the most important anthropogenic source of carbon dioxide, require several decades. No simple solution for this dilemma is in sight. In view of the fact that the world population is doubling every thirty years and the tremendous amounts of energy required to raise the standard of living in the developing countries, it is impossible to do without this massive and globally still-increasing consumption of coal, mineral oil and natural gas. All the more urgently then, both specialists and the public at large must realise the gravity of this problem. Methods for energy production ought to be developed that increase neither the level of atmospheric carbon dioxide nor the radiation burden of our environment.

The nitrogen cycle, certainly studied in less detail than the oxygen–carbon cycle, is a cause for serious concern too. The total amount of fixed nitrogen that man is adding to the soil, deliberately as fertiliser, unintentionally as combustion products and industrial exhaust, amounts according to even the most cautious of estimates to 20 to 30 per cent of all naturally fixed nitrogen. The production of fertilisers is likely to treble by the year 2000 and the 'unintentional fertilising' with combustion products is rapidly increasing, too. Nobody can reliably predict the consequences of such an increased addition of nitrogen compounds to the soil. If they are not decomposed, the concentration of toxic nitrogen compounds in subsoil water is likely to increase. If decomposed by micro-organisms, larger amounts of nitrous oxide (laughing gas), one of the products of such a process, will be liberated into the atmosphere. More nitrous oxide thus eventually enters the stratosphere and leads to a reduction in the atmospheric ozone concentration, which protects us from dangerous solar ultraviolet radiation, and which is being attacked by other products of our industrialised civilisation as well. There appears to be no chance of a simple solution in this case either. In view of the present food situation and the rapidly growing world population it is impossible to slow the rate of increase let alone decrease the amount of nitrogen fertiliser being used. Thus all the more urgently, ways have to be found to tackle the problem before it gets out of hand.

Many of the problems dealt with in this book describe scientific discoveries made only in the last few years. Much is still unclear, disputed or understood only in general terms. 'To ask what kind of information would be most useful ... is like asking a starving person to list the kinds of food he would be willing to eat,' declared the American geochemists R. M. Garrels and A. Lerman in a paper they presented to the Dahlem conference on Global Chemical Cycles and Their Alterations by Man. 'Almost any new data about the Earth would be welcome.'[2]

In this field of research there are hardly any figures known with the same degree of certainty as say the Earth's mean distance from the Sun or the atomic weight of hydrogen. In many cases one is entirely reliant on rough estimates, and the opinions of different specialists often differ by large amounts. When in 1970 Conrad C. Delwiche, an American biochemist, published results of the first budget calculations for the nitrogen cycle he commented that errors as large as a factor of ten might well be possible.[3] This means that the correct values might be ten times larger or even ten times smaller than those given in his paper. Within the years that have elapsed since then, he, and other scientists as well, have narrowed slightly this range of uncertainty. Nevertheless, when I had the chance to discuss this with him during the Dahlem conference mentioned above he said to me: 'You'll find the figures on the nitrogen budget in the conference folder. *Don't trust them!*'

This warning referred *to his own figures* as well as to those of some of his colleagues.

It goes without saying that theories and forecasts based upon such uncertainties can have only a tentative character. 'He who is willing to say what the final effects of such processes will be is wiser or braver than we are,' write the Americans P. Cloud and A. Gibor. 'Perhaps the effects will be self-limiting and self-correcting, although experience should warn us not to gamble on that ... If we want to ensure that the biosphere continues to exist over the long term, each new action must be matched with an effort to foresee its consequences throughout the ecosystem and to determine how they can be managed favourably or avoided.'[4]

Maybe further research will prove that some of the threats described in this book are not as serious as is suspected today. Likewise, however, one or another of the potential threats may prove worse than presently assumed. Today, man is, in principle, capable of severely disturbing many natural cycles. This is a fact from which we cannot escape. It appears inevitable that the extent of such potential disturbances keeps growing with the increase in the world's population and the rate of industrialisation. One simply cannot let everything take its own course until the day when scientists may be able to predict the consequences of our activities with greater reliability. Where doubt exists, precautions should be taken. Otherwise, we run the risk of overburdening our environment and causing damage which perhaps might be irreversible.

The topics dealt with in this book were thoroughly discussed at the Dahlem conference held in Berlin in November 1976. For five days, fifty geologists, meteorologists, oceanographers, chemists, ecologists, nuclear physicists as well as political economists and sociologists, from nine countries, attempted to define the problems, propose specific questions for further research and give preliminary recommendations on the basis of the present knowledge. I am indebted to the organising committee of this conference, in particular to its head, Dr Silke Bernhard, for giving me the opportunity to attend as an observer and to talk with several of the participants.

The discussions initiated during the Dahlem conference have been subsequently continued in further international and interdisciplinary meetings, one being the workshop on Carbon Dioxide, Climate and Society[5] held in February 1978 at Baden, Austria, by the International Institute for Applied System Analysis (IIASA). I am indebted to IIASA and in particular to the director of its energy project, Professor Wolf Häfele, for inviting me to this workshop. I gained new insight into this field and valuable material and ideas for this book from my participation in these two conferences.

I thank Professor E. Broda, University of Vienna, Professor C. E. Junge, Professor M. Schidlowski and Dr J. Hahn, Max-Planck-Institut für Chemie at Mainz, Professor K. E. Zimen, Hahn-Meitner-Institut für Kernforschung in Berlin, Professor H. Flohn, University of Bonn, Professor H. Lieth, University of Osnabrück, Dr H. D. Freyer, Kernforschungsanlage Jülich, Dr U. Hampicke, University of Essen, Professor B. Bolin, University of Stockholm, Professor H. Leibundgut, Federal Institute of Technology, Zürich (ETH), Professor W. S. Broecker, Lamont-Doherty Geological Observatory, Professor C. C. Delwiche, University of California, Professor R. M. Garrels and Professor F. T. Mackenzie, Northwestern University, Professor W. Reiners, Dartmouth College, Professor G. M. Woodwell, Marine Biological Center in Woods Hole, and Professor Stephen Schneider, National Center for Atmospheric Research at Boulder, for finding time to answer my questions and to read parts of the draft of this book. They have given valuable advice for the final version and for updating its English edition. I have to thank Dr B. Lötsch and Magister R. Göller of the Ludwig-Boltzmann-Institut für Umweltwissenschaften und Naturschutz, Vienna, Professor E. F. Brünig, University of Hamburg, and Dr E. T. Alexandersson, University of Uppsala, for providing instructive source material.

I am indebted to Dr P. Fabian of the Max-Planck-Institut für Aeronomie at Lindau and Dr W. A. Matthews, Alexander von Humboldt Fellow at the same institute, for their efforts to translate it into readable English. For the content of this book, and in particular the conclusions summarised in the last chapter, I bear exclusive responsibility.

Vienna, November 1979 GEORG BREUER

I

Oxygen and carbon dioxide – wherein lies the danger?

Are forests oxygen producers?

A full-grown tree has, as one can read in any literature on photosynthesis, a tremendous potential for producing oxygen. When it is about 115 years old, a beech tree exposes about 200 000 leaves with a total surface to 1200 square metres. During the course of one sunny day such a tree inhales 9400 litres of carbon dioxide to produce 12 kilograms of carbohydrate, thus liberating 9400 litres of oxygen. Through this mechanism 'about 45 000 litres of air are regenerated which is sufficient for the respiration of 2 to 3 people'.[1]

Similar figures can be found in many other papers on photosynthesis, and there is no doubt that these values are scientifically sound. Thus a reader not familiar with this subject might conclude that a mere accumulation of beech trees produces a *surplus* of oxygen. Yet, in fact, this is not true. Photosynthesis is only part of a closed natural cycle with respiration playing a reverse role.

Through photosynthesis green plants produce carbohydrates from carbon dioxide and water. In this process solar energy is taken up and stored in the organic matter, while the 'waste product' oxygen is released into the atmosphere. In respiration, the stored energy is used up: carbohydrates are 'burnt', consuming oxygen to form carbon dioxide and water – this is analogous to burning coal in a stove to utilise its stored energy.

Our beech tree needs about half the carbohydrate it produces to meet its own energy requirements. Through the respiration process of its leaves and roots, it consumes half of the oxygen it liberated, or to be more concise, it consumes an amount of oxygen equivalent to half the amount it produced. About another quarter is used up by respiring animals which feed on the short-lived products of the beech tree: caterpillars and beetles eating some of the leaves, animals in the soil and microorganisms living on autumn foliage that has fallen to the ground, while squirrels and wild boars eat the beech seeds, etc. The remaining quarter of the oxygen liberated by our beech tree is consumed when its wood – or the paper which is eventually produced from it – is burnt and when its stump, roots and branches decay, i.e. are decomposed by small animals and microorganisms that consume oxygen.

Of course the relative amounts entering the budget do vary. However, whatever the relative amounts might be, in the end the budget always balances. *When the total organic substance produced by the beech tree during its lifetime is decomposed, the total amount of oxygen consumed is equivalent to that liberated when it was alive.*

If our beech tree grows at the edge of a swamp its foliage, branches and finally the whole trunk might sink into the muddy

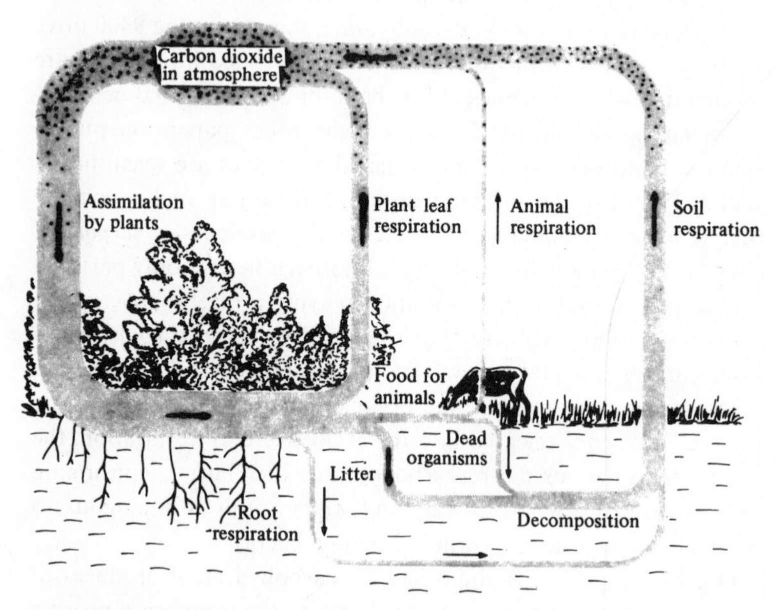

Figure 1 Schematic representation of the atmospheric–biospheric carbon cycle. The widths of the pathways are roughly proportional to the quantities involved. In middle and high latitudes, soil respiration, i.e. the sum of root respiration of plants and respiration of microorganisms and small animals in the soil, can amount to almost seventy per cent of the total respiration rate. Plant leaf respiration makes up about twenty-five per cent, while animal respiration (including human breathing) account for five per cent. In the tropics the fraction of soil respiration is smaller. There, the humus layer is in general thin since organic matter decays at a much faster rate and mostly on top of the soil. (From B. Bolin, *Scientific American*, September 1970, p. 126.)

ground, where, in the absence of oxygen, they eventually carbonise. Only under such circumstances will an oxygen surplus be produced by this tree that remains in the atmosphere until the carbon, through erosion or mining, gets back to the surface and can thus be oxidised. This process may require thousands or even millions of years. In this case the natural cycle becomes closed with a long delay through geological rather than biological processes.

Put simply, we human beings breathe the oxygen which was produced by the crops on which we live or which nourished the animals, the products of which we eat. Of course, we cannot take this literally, since the oxygen molecules cannot be traced back precisely to the carbohydrate molecules whose formation liberated them. But quantitatively there is agreement; while breathing, we use up exactly as much oxygen as is necessary to burn the food we eat. On the other hand the amount of organic matter (carbohydrates) available for oxidation cannot exceed that previously produced by green plants. In the present biological environment the production and usage of oxygen and carbon dioxide are essentially balanced.

However, the decay rate of organic matter differs according to the pathway of oxidation. While the carbon fixed through photosynthesis remains in the organic substance, the liberated oxygen remains in the atmosphere. The fact that there exists biologically fixed carbon which has not yet been reoxidised is the reason for the presence of oxygen in our atmosphere. In this context scientists talk about different reservoirs of carbon with different residence times (the period before the carbon is reoxidised).

From the first reservoir, the plants themselves take the carbohydrates necessary to meet their own energy requirements. Here the residence time of the carbon varies between fractions of a second and a day, since plants also respire during the night when, due to the absence of sunlight, no photosynthesis occurs. The daily variation in this reservoir can clearly be recognised by measuring the composition of the air in a forest area. As soon as the Sun rises and photosynthesis begins, one can usually measure a decrease in the carbon dioxide concentration. In the late sixties when the average abundance of carbon dioxide was 320 parts per million (1 ppm stands for one molecule of carbon dioxide among one million molecules of air) or 0.032 per cent, 305 ppm of carbon dioxide were measured during sunny afternoons close to the tree-tops in a forest. However, on fairly still nights up to 400 ppm were observed close to the ground in the forest where decomposition processes continue both day and night.[2]

Such a considerable increase in the carbon dioxide content must be accompanied by an equivalent decrease in the oxygen content and vice versa. However, these localised oxygen variations are so small when compared with the huge oxygen amounts present in the atmosphere that they are beyond the detection limits of present-day instruments.

In the second reservoir the residence time-scale of the carbon is of the order of weeks to years. Here we find short-lived biological products (provided they do not contribute to humus formation): crops, grass, flowers, foliage, and needles of trees, and all those parts in which plants store energetic substances for their offspring or for the next growth period – fruits, seeds, bulbs and so on. Most animals can also be placed in this reservoir, although the total mass of all animals that live on land amounts to only one thousandth of that of the land vegetation.

In moderate and high latitudes the content of the second res-

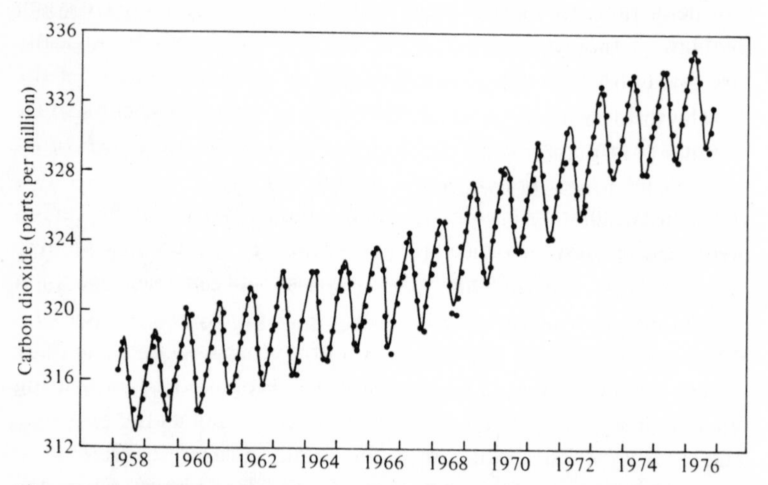

Figure 2 Variations of atmospheric carbon dioxide concentration measured by C. D. Keeling and collaborators at Mauna Loa Observatory, Hawaii. The seasonal oscillations are caused by the seasonal changes in vegetation in the northern hemisphere. The dots indicate monthly averages. (Graph from G. M. Woodwell, *Scientific American*, January 1978, p. 37).

ervoir is subject to pronounced seasonal variations. This is clearly reflected by measurements of atmospheric carbon dioxide made by American scientists since 1957 at Mauna Loa Observatory, Hawaii. These data show a gradual increase in the average carbon dioxide content over the years superimposed with a distinct annual wave-like variation. A similar trend was found by Swedish meteorologists who have been measuring carbon dioxide aboard aircraft regularly since 1963.[3]

According to estimates made by the Swedish meteorologist Bert Bolin the atmosphere north of 30° N loses nearly 3 per cent of its carbon dioxide content during the summer months from April to September. This corresponds to about 4 thousand million tonnes of carbon. In the near-surface air over the continents of the northern hemisphere, the average carbon dioxide mixing ratios varied in the mid-sixties from more than 330 ppm in April to less than 310 ppm during September. Those variations are not confined to regions covered by forests, since winds cause effective mixing throughout the whole of the northern hemisphere. In Hawaii, far away from the continents, and with tropical latitude, the average variation between spring and autumn still amounts to 5 ppm. In the southern hemisphere, with only a small fraction of land surfaces at middle latitudes, these annual variations are less pronounced. Nevertheless, they are clearly recognisable even at the South Pole.

The third reservoir contains the wood of living trees and humus, remnants of plants being decomposed by a vast number of microorganisms, fungi, worms, mites, insects, and other small creatures. Some long-living species of animals and the human race (or better, the carbon contained in them) can also be placed in this third storage level which holds more than 90 per cent of the living substance, and nearly all of the decomposing substance, on land. This carbon is fixed for tens to hundreds, sometimes even thousands of years. It amounts to at least twice to three times the carbon stored in atmospheric carbon dioxide (see figure 5 on page 18).

By far the greatest reservoir is the fourth which contains the remnants of plants and animals which, being buried in sediments (deposits of clays and sands), are taking part in long-term geological cycles. The sediments gradually solidify to form rocks, which either become involved in mountain formation or get buried in the floor of the oceans. In the first case, as the mountains gradually get weathered, the carbon embedded in the rocks is exposed to the atmosphere and thus oxidises. In the second case, the sediments penetrate to deeper layers where melting and various chemical processes occur. Here, too, the carbon usually becomes oxidised, eventually returning to the atmosphere as volcanic carbon dioxide.

In this reservoir the carbon remains as a rule for several hundred million years. The total mass of carbon in this fourth reservoir is about one thousand times that of the total carbon in all living and decomposing organisms on the continents and in the oceans. Most of it is dispersed in the ocean sediments. Only about one or two thousandths constitute deposits of fossil fuel worth mining. In early geological times, when large coal deposits were formed, this storage level was fed to some extent from the land too. At present however, according to experts, it is almost exclusively supplied by the oceans.[4]

Of course it is not always exactly the same carbon that is stored in the respective reservoirs. A forest does not always consist of the same trees: old ones fall and rot away while young ones grow. But as long as the total mass of wood and humus of a forest varies only slightly, as is generally the case under natural conditions, the total mass of carbon fixed in the different reservoirs remains more or less constant. Thus an equivalent constant amount of oxygen, liberated through carbohydrate formation, remains in the atmosphere.

The composition of the atmosphere does not depend on the rate at which the carbon passes through a reservoir. The only important figure is the total mass fixed within that reservoir. In a tropical rain forest for instance, the production and consumption of oxygen and carbon dioxide per unit area per year amounts to about twice that observed in a mixed forest at moderate latitudes. The mass of carbon fixed in the wood and humus, however, is about the same in both cases with respect to the unit area. Hence the equivalent amounts of oxygen remaining in the atmosphere are about equal for both forest types.

Neither a primeval forest nor a cultivated forest where clearing is balanced by immediate reforestation can contribute to restoring the oxygen used up somewhere else (geological processes, industry etc.).

In this sense forests are *not* oxygen generators. Only where a permanent increase of the living substance (the biomass) and the humus layer is achieved, for instance by planting trees in deforested areas, will the composition of the atmosphere be altered. That is, more carbon dioxide is taken up from the air than is liberated by the processes of life. The amount of carbon retained in the third reservoir increases. Consequently, through photosynthesis, more oxygen is liberated than consumed by respiration and decay processes, until the new growing forest has reached maturity and the cycle is balanced at a new level.

Unfortunately the reverse process of forest destruction is more common nowadays, leading to a decrease of atmospheric oxygen and a corresponding increase in carbon dioxide levels. This happens even if the deforested areas are successfully converted to farmland, since the farm products are short-lived and thus enter the second carbon reservoir instead of the third. Most of these products are consumed within a year, which means they are reconverted into carbon dioxide and water consuming all the oxygen liberated during their period of growth. The amount of carbon fixed in crops varies between 0.2 and 6.0 kilograms per square metre, with a global average of 0.5 kilograms per square metre for farmland, with strong seasonal variations. The amount of carbon stored in the wood of living trees of a forest is about ten to one hundred times this and can be even more.[5] Moreover, through the conversion of a forest to farmland, and worse to eroded land or desert, the humus layer is destroyed.[6] Such humus destruction is not part of a short-term biological cycle. Rather, it leads to a single massive consumption of oxygen and a corresponding increase in the level of atmospheric carbon dioxide.

Forest destruction and carbon dioxide increase

One third of the global land area, about 50 million square kilometres, is covered by forest, or at least it was in the middle of this century. How far would the forests have extended *without* man's intervention upon nature? Half the global land area? Two thirds? Or even more?

During an interval in the Dahlem conference I addressed this question to Professor George M. Woodwell, Director of the Ecosystems Center of the Marine Biological Laboratory in Woods Hole, Massachusetts. 'Well', he replied, 'the potential forest areas are well known. As a rough guess I would say at least half of the land area. Nearly all of Europe would be covered by forests, the Mediterranean countries, most of the Middle East, India, China, the east of the USA, nearly all of Canada, and large parts of Latin America. But there are parts of America, Africa and Asia which would be savannas, grasslands or deserts even without human interference.'

Certainly there has been a natural change in the vegetation since the end of the last ice age, some ten thousand years ago. The forests have been spreading northwards, and the southern boundary of the tundra has been receding towards the North Pole. The climate in the subtropics has become considerably dryer in the last four thousand years. Thus the desert belts would probably have spread solely due to climatic changes. However, scientists are convinced that this development has been considerably accelerated by human activities.

The extension of formerly wooded areas converted to desert by man was estimated in 1957 by the German forest expert Johannes Weck to amount to 4.5 million square kilometres. Mohammed Kassas of the University of Cairo suggested in 1976 that the desert areas resulting from man's activities nearly equal the total farmland area on Earth, which is about 14 million square kilometres; one third of this agricultural land, about 5 million square kilometres, will become desert before the end of this century if present trends persist, according to a prognosis made by experts at the United Nations conference held in Nairobi, September 1977.[7]

One can assume that most of the present farm and pasture land was created by the clearing of forests (See table 2 on pages 172–3 for figures on areas). Likewise the tropical savannas do not represent the original vegetation of these areas. As the German ecologist Sigmund Rehm explains, these 'nonproductive grass-deserts' developed from forests due to the wrong exploitation of the soil.[8] In the same sense botanists of the University of Mexico City talk about 'anthropogeneous (man-made) savannas and savanna woodlands' in Mexico, South America, Asia and Africa,[9] and nearly everywhere in the Mediterranean countries where 'macchie' (shrub and bush) is found today, genuine forests existed before Greek and Roman times.

The cutting of those forests which covered the present farmland areas alone has meant that the amount of carbon stored in living plants on land has been reduced by about 170 thousand million tonnes. This reduction, as estimated by the American ecologist Robert H. Whittaker, had its onset with the beginning of human civilisation.[10] It corresponds to about 25 per cent of the carbon stored in atmospheric carbon dioxide. Moreover, the carbon from all those forests converted to savanna, scrub and desert was oxidised, and as a consequence most of the humus was also destroyed. Thus one can assume that during the last few thousand years, through human activities, an amount of carbon probably not much less than the amount present in the atmosphere was transformed into carbon dioxide.

Provided these processes occur slowly, the additional carbon dioxide produced is absorbed by the oceans and thus no considerable accumulation in the atmosphere is likely to occur. During the last two hundred years, however, the rate of forest destruction has been increasing in an alarming manner. Thus ecologists and biologists, almost unanimously, are of opinion that the increase in atmospheric carbon dioxide observed during the last decades is not only due to increasing consumption of fossil fuel but also due to the destruction of forests and humus.

Unfortunately hard figures on global forest areas and their annual changes are very scarce. The statistics of the World Food and Agricultural Organisation (FAO) are insufficient, says George Woodwell. The inventory of the International Biological Programme constitutes only a first and hence unreliable estimate. There are however many individual reports, mostly describing the situation in tropical countries, which yield an alarming mosaic.

The huge Amazon Orinoco basin, with about 4 million square kilometres, constitutes the largest coherent forest area on Earth. Brazilian government plans envisage that most of this primeval forest be converted to farm land or cultivated secondary forest by the end of this century. Such conversion, disregarding local conditions is very likely to end in disaster. (E. F. Brünig, Institute of World Forestry, University of Hamburg, FRG.)[11]

Figure 3 Clearing by burning in the Amazon basin. The plans of the Brazilian government to convert the tropical rain forest in this area into farmland, pasture land and cultivated forest are likely to end in ecological disaster (Photograph: F. Müller-May, *Bild der Wissenschaft*.)

In 1910, 60 per cent of the area of the state of São Paulo (250 000 km^2) was covered with virgin forest; by 1950, only 20 per cent was still so forested and clearing was continuing at a slightly slower rate. In the state of Parana, in the period from 1953 through 1963, an average of over 3 per cent of the 1953 forest area of 65 000 km^2 was cleared each year with the burning of the felled trees, causing an annual economic loss estimated at about 2.5 thousand million US dollars ... Reforestation in Parana is estimated at 10 per cent of the cuttings and only in very recent years has reforestation reached 20 per cent of the cuttings in São Paulo. (J. A. S. Adams and others, Geophysical Institute, University of São Paulo, Brazil.)[12]

Only the fall of the foliage evenly distributed throughout the year nourishes the vegetation and maintains the biological equilibrium of the primeval forest. As soon as it gets cut down by man, the soils are quickly exhausted and even become totally barren. (H. Grabert, Geologisches Landesamt, Krefeld.)[13]

In virgin tropical forests the incoming solar radiation is absorbed by the canopy formed by the leaves. Virtually no radiation penetrates to the ground and heats it. The evaporation of water is almost exclusively regulated through the transpiration of the trees. If the forest is cut a great deal of solar radiation reaches the ground. Hence the evaporation from the ground is greatly enhanced. The top soil layer rapidly dries out, endangering the water supply for the plants. In heavy rain the impact of the rain drops hitting the top soil leaves a hard crust after drying that bars the oxygen supply from the atmosphere, so vitally important for root respiration. For traditional farming methods these consequences are not so serious. The introduction of the plough and the rationalisation of cropping (mono-cultures) as customary in mid-latitude farming, however, cause the soil to be exposed or only barely covered by plants for longer periods. Thus the likelihood of harmful structural changes is greatly increased by these methods compared to the former 'primitive' farming methods. (S. Rehm, Institute for Tropical and Subtropical Agriculture, University of Göttingen.)[14]

Over 95 per cent of households in developing countries where woodfuel is readily available, use it as a primary source of energy. Similarly in these countries industries such as brick and ceramic manufacture, fish and tobacco curing, cassava meal, rubber and sugar production, tin smelting and steel manufacture all use woodfuel to a varying degree. In areas where good wood is readily available, only dead branches are collected, but where shortages occur, young saplings are taken, and eventually mature trees are felled especially by the 'urban' charcoal producer. The demand on woodlands and forests has led to a denudation of the land in certain areas. The removal of the vegetation cover has also increased soil erosion which, at least, causes a loss of topsoil and can at worst, result in massive flooding in river valleys and delta regions. Likewise the continual cutting of fuelwood in the southern Sahara/Sahel region has facilitated the rapid shift southwards of the desert. (K. Openshaw, International Forestry Science Consultancy, Penicuik, Scotland, who studied wood consumption in developing countries on behalf of the FAO.)[15]

The lower ranges of the Himalayas are today a sorry picture of man's inhumanity to nature. From Kashmir to Burma, a vast tract of land has been laid barren and desolate, where tigers once lived amid lush forests. (A. Agarwal, science correspondent of the *Indian Express*.)[16]

In eastern central Mexico a rapidly expanding rural population is clearing the mountain forests at a frightening rate while retaining traditional farming methods. Recently the necessary period to leave the land fallow has been progressively shortened and this involves increasingly larger areas that were formerly covered by forests. This uncontrolled felling is likely to bring up severe threats to the future of Mexico. Increased erosion, the lowering of the water table, and dust storms are likely consequences. (H. Ern, University of Constance.)[17]

If an increasing population demands shorter regeneration periods for forests and longer farming times a noticeable decline in harvest output is the only consequence. On an exhausted soil forest regeneration becomes weaker, and perennial grasses gradually dominate. Vast areas of unusable land covered by grasses such as Alang-Alang are the result. In Indonesia alone, these cover more than 15 million hectares, more than the total area used for cropping, tree cultivation and pastures put together. (S. Rehm.)[18]

With the present rate of destruction of the tropical rain forests throughout the world, there is a great danger of the mass extinction of thousands of species. This is due to the simple fact that primary tree species from the tropical rain forests are incapable of recolonizing large areas opened to intensive or extensive agriculture. (A. Gomez-Pompa and others, University of Mexico City.)[19]

In Costa Rica, the Government Colonisation Institute can clear a hectare in only two hours; in Kalimantan (Borneo, Indonesia), the International Timber Corporation of Indonesia is logging at the rate of 60 000 cubic metres a month, and expects to achieve 80 000 m^3 a month soon. If current attitudes and practices persist, all the world's tropical rain forest, apart from a few scientific mementos, will be destroyed within 20 to 30 years. (Robert Allen, International Union for the Conservation of Nature and Natural Resources, Morges, Switzerland.)[20]

At the beginning of this decade the tropical forests covered about 20 million km^2. The growing stock of this area contained nearly 10 billion tons of biomass (dry matter), more than half of this in the tropical rain forests. At the present annual rate of destruction of 0.3 million km^2 of tropical rain forest, natural forest on easily accessible land will have disappeared from this zone by the end of the century. In this process at least 300 thousand million tons of dry matter will have been burnt or decomposed, using 400 thousand million tons of oxygen (0.03 per cent of the oxygen stored in the atmosphere) and releasing 550 thousand million tons of carbon dioxide (equal to about one quarter of the atmospheric store) into the atmosphere. 10 to 20 per cent of the released carbon dioxide will eventually be fixed again by the substitute vegetation, 40 per cent will be buffered by the oceans, but approximately 40 per cent will remain in the atmosphere and increase its carbon dioxide content by 10 per cent above its present level. (E. F. Brünig.)[21]

In addition large quantities of carbon dioxide are liberated through humus destruction following deforestation. Moreover, it is not merely tropical rain forests that are being destroyed but others as well, and it is not only in developing countries but also in highly industrialised states such as Canada and the Soviet Union. In particular, the opening of Siberia is associated with the destruction of considerable forest reserves.[22]

'Do you know of any developments going the other way, i.e. forest augmentation?' I asked Professor Woodwell.

'Yes, I do,' he replied. 'In Europe marginal agricultural land and pasture are continuously being re-transformed into forests. There were such tendencies in the east of the USA too, but it seems that this trend has stopped now. There are reports about big afforestation programmes in China, but little is known about their success; we do not know whether the newly planted forests are developing well. Moreover there may still be a slow northward movement of the northern boundary of forests, so a marginal increase in the forests is possible there. Compared with the massive forest destruction in tropical countries, such changes are small however. Globally, I am sure that forests as a whole are diminishing and that the biomass is decreasing.'[23]

In developing countries forests are often burnt down, i.e. immediately converted into carbon dioxide, in order to clear land for primitive farming. About two thirds of the world's wood harvest is used for wood-fuel.[24] A considerable fraction of the remaining third is used for the production of paper and other products which are burnt or rot within a short period of time. The average lifetime of timber, railway sleepers and wood used in carpentry is likely to be about the same as that of wood under natural conditions. Only a small number of wooden artefacts are removed from the natural cycle for centuries or maybe thousands of years. The amount of carbon stored in these for long periods is however so small that it can certainly be neglected with respect to the total budget.

Erosion and oxygen loss

Forest destruction does not only result in an increase of atmospheric carbon dioxide but there are many other consequences which, at least in the short-term, constitute an even greater impact. Forests are like huge sponges, capable of storing precipitating water which is then gradually released through both evaporation and slow runoff in the subsoil and surface waters. Forests act as dust filters, they protect against landslides and avalanches, and even in narrow strips can provide protection against wind and noise. All these beneficial effects which in one way or another influence the atmosphere, are lost when forests are destroyed.

Whenever forests are cut down for timber production, a sustained yield can only be achieved when deforestation and reforestation proceed hand in hand. In most cases, however, especially in the developing countries, unrestrained exploitation is the rule. The culprits are not only the poor and ignorant peasants short of wood-fuel, but also the big timber companies which are usually neither familiar with the local conditions, nor in contact with the population.[25]

In India there was a widespread movement of civil disobedience among the inhabitants of the southern slopes of the Himalayas against the reckless methods of the timber industry. In order to prevent further tree cutting, villagers embraced the trees which were to be felled. This nonviolent resistance, in true Gandhi spirit often forced troops of wood-cutters to withdraw.[26]

In many cases forests are cleared to create new farmland in order to increase food production, but this is often not achieved, particularly in steeply sloping regions. Subsequent changes in the water budget, flooding, soil erosion and landslides result in a *decrease* in total food production in that region despite the enlarged farming area.

When mountains are bare of forests, every heavy thunderstorm can cause a disastrous flooding. 'The principal cause of the recent floods on the Indian sub-continent was the removal of the tree cover in the catchment areas,' writes K. Openshaw.[27] A. Agarwal reports disastrous flooding of the Alakananda river following total deforestation in that area in 1970. A whole village and several busloads of tourists were swept away by the floods and five big road bridges, thirteen suspension bridges and two wooden bridges were destroyed within a matter of a few hours. Further devastation occurred when the huge amount of silt carried along with the water was deposited in the lowland and blocked about 10 kilometres of an important irrigation channel.[28] A report presented at the Eighth World Forestry Congress, held in Jakarta in 1978, states that the Mangla reservoir in Pakistan receives about 100 million tonnes of sediment each year, four fifths from the deforested catchment of the Jhelum river. The reservoir was built to last for one hundred years, but most of its capacity could well be lost within half that time.[29]

In the early seventies the pine forests covering the northern slopes of Pico de Orizaba in Mexico were cleared up to about 3500 metres above sea level, in order to gain new farmland for corn and potato production. It was as early as 1974 that the geographer Klaus Heine from Bonn found an erosion furrow which had developed there to a length of 350 metres, 5 to 30 metres wide and 4 to 10 metres in depth. 'Farming on these slopes will become impossible within a few years' time,' he predicted.[30]

It has always been like this. Since the early days of farming man has always over-used his environment, exploiting the land until it will produce no more. This caused the decline of the Sumerian empire in the Middle East five thousand years ago, and many many other flourishing cultures have met the same fate. In ancient times men like Plato and Ovid lamented because of what we would call the destruction of the environment. In America, long before its 'discovery' by the white man, the Aztec city of Teotihuacan was surrounded by a belt of eroded land.[31]

Neither the Sumerians, nor the ancient Greeks nor the Aztecs

Figure 4 Deep erosion furrows are resulting from the removal of the protective forest cover. In the rainy season, the fertile topsoil layer, which is generally thin in the tropics, is rapidly carried away. (Photograph: F. Müller-May, *Bild der Wissenschaft*.)

could foresee the consequences of their farming habits, and even today the poor landless peasants in Mexico or in the Sahel often do not know that they threaten the roots of their own existence by cutting down the forests. But we, the inhabitants of the highly industrialised countries, *do* know the consequences. We also know how to produce enough food without totally destroying the natural environment. This knowledge is very rarely reflected in our developmental aid however. Instead we provide the Indians with bulldozers and chain-saws to strip the Himalayas of forests even faster than the Romans and the Venetians did in the mountains of Dalmatia.

Forty kilometres north of Rome there is a little lake called Lago di Monterosi. During the second century BC the Romans built the Via Cassia through this region thereby opening it up to human civilisation. Today's geologists have examined the material deposited at the bottom of this lake which exactly reflects the consequences of human activities since then. Before the Via Cassia was built the average erosion rate was about two to three centimetres in each thousand years; since the 'cultivation' of the region it amounts to about twenty centimetres per thousand years.[32]

Erosion measurements made in the state of Mississippi yielded average annual loss rates of top soil amounting to about 10 to 50 kilograms per hectare (kg/ha) for forests, about 800 kg/ha for pasture land, and the tremendous rate of 10 tonnes/ha for farmland. These results were confirmed by other scientists in the USA and Europe. For the total area of the United States the American geologist Sheldon Judson estimates the present erosion rate to be at least three times that of the age before civilisation.[33]

Before agriculture was introduced, the total mass of sand, shingle and humus carried to the oceans by all rivers on Earth was, according to Judson's calculations, about 9 thousand million tonnes a year. His colleague B. Gregor, using other methods for these calculations, came up with 10 to 11 thousand million tonnes per year.[34] Both scientists agree that before the middle of this century these numbers had already increased to about 24 thousand million tonnes per year. Through the massive destruction of tropical forests erosion is being further increased. According to one of the papers presented at the Dahlem conference, giving preliminary data, the present rate is already estimated to be four times the average of the last 700 million years.[35]

With regard to the composition of the atmosphere, these findings are more important than they appear at first sight. Increased soil erosion means that every year more rocks are denuded, thus becoming exposed to atmospheric oxygen. Hence increasing amounts of oxidisable substances contained in rocks, in particular iron, carbon and sulphur, react with atmospheric oxygen, thereby increasing the geological oxygen loss. In the very long run, that is millions of years, this might lead to a noticeable reduction in the atmospheric oxygen level.

Are algae oxygen producers?

In the oceans, photosynthesis is nearly exclusively sustained by myriads of microscopically small algae. The contribution of large multicellular plants is negligibly small. All other organisms in the ocean, from the tiny zoo-plankton to the huge whales, live directly or indirectly on the organic substance produced by this tremendous quantity of unicellular algae; and they use up the oxygen liberated through the photosynthesis of this substance. If whales and the few other sea creatures with lungs, and the catching of fish by man, are disregarded, we are dealing with a natural cycle of dissolved gases which never enter the atmosphere.

The average life-span of the algae, and of the little animals feeding on them, varies from a few weeks to a few months, while that of the larger animals at the other end of the food chain does not exceed some tens of years. The total mass of sea plants amounts to about one thousandth of the biomass on land. The mass of all animals nearly equals that of all plants in the ocean and hence equals that of all animals on the continents, whereas on land, animals make up only about one thousandth of the total biomass (see page 4).[36]

In the oceans, decay of organic substances mainly occurs through the food chains of animals. In addition, there is some nonbiological direct oxidation of remnants of dead microorganisms or dissolved substances. Compared to the decay processes on land,

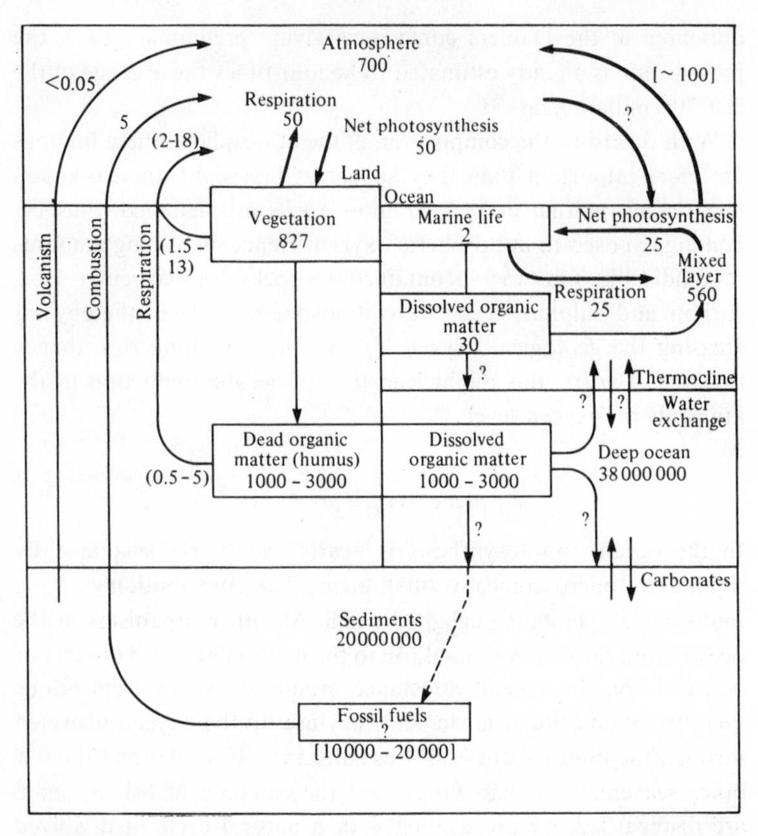

Figure 5 Budget of the global carbon cycle according to G. M. Woodwell. Units for the figures referring to the different reservoirs are in thousand million tonnes of carbon; units for the transfer rates are in thousand million tonnes of carbon per year. Figures in brackets represent estimates from other sources. The thermocline is the boundary between the surface layers, subject to seasonal temperature variations, and the permanently cold deep waters. There is only very little water exchange across the thermocline. For further explanation of the carbon cycle see part 3 of this book. (From G. M. Woodwell, *Scientific American*, January 1978, p. 9.)

bacteria and fungi are much less important. Some part of every sea organism that dies, as well as faecal matter, sinks to the bottom of the sea where it nourishes a large number of animal species. The total mass of these faeces, dead algae and dead animals floating and gradually sinking to greater depth, is enormous; it amounts to more than a thousand times that of all living organisms in the oceans. It is comparable to, or even greater than, the total mass of humus on land. The average residence time of carbon in this reservoir is estimated to be about three thousand years. It is hence considerably longer than the residence time in the third reservoir on the continents.

In the fifties people still believed that the huge volume of the world's oceans would produce huge amounts of organic matter. The contribution of the oceans to photosynthesis of all plants was estimated to amount to about 60 to 80 per cent. More recent research has shown however that the fertility of the oceans is far smaller, its productivity reaching merely one fifth to one quarter of the average productivity on the continents. Thus, since the ocean surface is about twice that of the continents, the maritime contribution to the world production of organic substance is estimated to be only about one third.[37]

The low fertility of the oceans results from the way the essential nutrients circulate in the water. It is not only the carbon compounds which sink with the dead organisms, but also nitrogen and phosphorus compounds, all gradually reaching depths never penetrated by sunlight and thus hostile to any growth of green plants. In the absence of green plants and hence photosynthesis, there is no new formation of organic substance. Thus there is no means by which the nitrogen and phosphorus compounds can be fed back into the biological cycles. Instead these essential elements enrich the depths of the oceans while there is often a deficiency in the surface waters where they would be needed for plant growth.

For this reason, the production of organic substance per unit area for most of the open sea is small, hardly exceeding that of desert areas and smaller than in the tundra. Thus one can classify the oceans as huge 'water deserts' with occasional fertile oases mostly occurring along the shorelines where productivity can reach that of the tropical forests. These are areas with upwelling deep water rich in nutrients, such as along the Antarctic coasts, the Humboldt current along the Peruvian coast, or estuaries and lagoons which carry nutrients of continental origin.

A small fraction of the organic substance sinking to the bottom of the ocean, according to Garrels and his colleagues, about 1.5 thousandth of the total production, gets buried in sediments.[38]

About one third of this amount is subjected to bacterial decay, the remainder enters the fourth reservoir (see page 5). The equivalent amount of oxygen liberated through photosynthesis of these carbon compounds remains in the atmosphere as 'surplus' until the carbon buried in the sediments is oxidised through geological processes, usually after hundreds of millions of years. From this point of view the algae do indeed provide oxygen. It is this one thousandth of their oxygen production that is not recycled, amounting to about a hundred million tonnes per year, which provides the reserves compensating for the geological loss of oxygen by the oxidation of iron, sulphur and volcanic gases. Throughout the hundreds of millions of years of the Earth's history, it has for the most part been these tiny annual 'surpluses' from which the huge reservoir of atmospheric oxygen has been built up.

Where ocean pollution from oil and other chemicals leads to a grave disturbance in the maritime life, there is the threat that many coastal areas and even complete seas like the Baltic will be transformed into man-made 'water deserts'. These developments have, regrettable and dangerous as they are, very little impact on the composition of the atmosphere, since the total biomass of marine organisms is small. Compared to the oxygen consumption by technical processes and forest destruction, amounting to some thousand million tonnes per year, the 'surplus' produced by algae is hardly more than a drop in the bucket. Even if man-made pollution is going to reduce this amount by several million tonnes, this will hardly have any impact upon the atmosphere in the next decades or even centuries.

Is atmospheric oxygen endangered?

'How did you happen to choose this field?' I was asked by Professor Manfred Schidlowski, a geochemist heading the research group studying the 'History of the Atmosphere' at the Max-Planck-Institut für Chemie, when I visited him in his laboratory at Mainz, while preparing this book.

'As a matter of fact it began by a mistake', I confessed.

'A mistake, how?'

'I was of the opinion that atmospheric oxygen is endangered', I replied. 'In an earlier book dealing with energy problems I had raised the argument that in the future there might not be a sufficient number of plants on Earth to regenerate the oxygen in the air.[39] Now I wanted to pursue this problem in more detail.'

Before our conversation had reached this point I had already been questioning Professor Schidlowski for a couple of hours about the activities of his research group. Now, while lunching together, our roles changed: he asked the questions and I answered. I told him that I had often found estimates in the scientific literature indicating that, in many industrialised countries, oxygen consumption for industrial purposes does by far exceed the oxygen regeneration by the plant cover of these regions.[40]

'The figures were certainly correct,' I said, 'but they led me to reason incorrectly. To start with, I was of the opinion that industrialised countries have to import oxygen in a sense from those areas where a surplus is produced, hence from the oceans. These imports, as I wrote in my book, are achieved easily without any costs or energy demands, by the winds. However, what will happen if on the one hand technical oxygen consumption continues to increase, while on the other hand more and more algae are destroyed by oceanic pollution? I had however overestimated the importance of the oxygen surplus generated in the oceans, and moreover, I had underestimated the tremendous amount of oxygen stored in the atmospheric reservoir.'

'But you were probably aware of the fact that even if all known and estimated reserves of fossil fuel are burnt, atmospheric oxygen will only be depleted by less than 3 per cent?'[41]

'Yes, but I saw the main problem as lying elsewhere. Oxygen is a product of green plants. It is continuously regenerated by them. Man is increasingly destroying plants, cutting down forests, creating deserts, and polluting the oceans with oil and chemicals, so that along many coasts life can hardly exist any more. I thought, if there are less plants on Earth, perhaps two thirds or half as compared to undisturbed natural conditions, then there must be consequences with respect to the oxygen abundance in the atmosphere.'

In this concern, I continued, I was by no means alone. Since the mid-sixties many well-known scientists had been expressing concern about atmospheric oxygen. I did not have to explain this in detail to my partner in the conversation for he was familiar with this. I would however like to quote some of this material for information of the readers of this book. In 1966 the highly regarded American scientists L. V. Berkner and L. C. Marshall, authors of the fundamental work about the evolution of atmospheric oxygen, wrote a memorandum dealing with the threat of a sudden drastic reduction in the oxygen concentration in the air. Although this paper was not published it has been quoted by other authors.[42] In 1967, during the conference of the American Association for the Advancement of Science, the American ecologist C. Cole LaMont presented a paper concluding that, because of oceanic pollution, three quarters of our oxygen supply might be endangered.[43] At other scientific conferences this prognosis has repeatedly been quoted by other scientists, and it has also found an enormous response in the media. In Europe too there has been concern about this problem. Newspapers with a large circulation reported it, and Th. Keller of the Federal Institute for Forestry Research at Birmensdorf, Switzerland, even suggested the introduction of a tax on oxygen consumption.[44]

'And how did you finally discover that you were on the wrong track?' asked Professor Schidlowski.

'Well, this was not so simple. First of all, while studying the literature, I certainly also found other opinions. Among them was a reply to Cole LaMont by the marine biologist John Ryther and, in particular, a short article by your colleague, Wallace S. Broecker, categorically stating that the atmospheric oxygen reserves are practically unlimited.[45] But how could I decide whether Broecker was right and Cole LaMont was wrong? It is not at all unusual for scientists to defend conflicting theories and often in retrospect analysis reveals that there is some truth on both sides. Thus I intended to give a balanced presentation to the readers of my book: a problem of general interest about which there is disagreement within the scientific community.'

I delved deep into the material, I continued. I wanted to understand the arguments of both sides as best I could, in order to be able to explain why some scientists do and other do not see oxygen as being endangered. For weeks I read, summarised, calculated, compared figures and arguments. And then one day, quite suddenly, I realised that the concern about the oxygen was indeed based upon an error.

Above all else there was the simple comparison of the orders of magnitude. The total amount of carbon in all living and decaying substances on land and in the oceans amounts to 7 billion tonnes at most. Assuming that through deforestation and oceanic pollution, half of all living beings are destroyed, an already fairly pessimistic scenario, amounting to 3.5 billion tonnes of carbon, an equivalent of about 9 billion tonnes of oxygen would be required to oxidise this amount. But this is less than 1 per cent of the 1200 billion tonnes of the oxygen presently in the atmosphere, and hence is a negligibly small amount. The carbon dioxide content of the atmosphere, however, would thus be drastically increased.

The second and yet very important point was that any permanent reduction of photosynthetic organisms must necessarily be accompanied by a corresponding reduction in the biological oxygen consumption. It is indeed possible that plants produce more carbohydrates than the total used by the respiration of all animals and microorganisms. This leads to an enrichment of oxygen in the atmosphere – a process which took place during the Earth's geological history. It is impossible, however, for respiring creatures to consume more carbohydrates than have been produced. They cannot use up more oxygen than had been liberated in the production of these carbohydrates.

'To ask the question "What would happen if all plants were destroyed and thus photosynthesis were terminated?" is only of interest for geological models,' I added. 'In all events, the population would starve to death in a world without plants long before anybody would suffer from a lack of oxygen.'

'Such models certainly exist,' Professor Schidlowski added. 'When all life on Earth is terminated the oxygen would gradually be used up oxidising iron and other elements that are exposed through erosion processes. According to estimates made by Holland, this process would take about three million years, and this figure was also used in model calculations made by our own group.[46] With respect to the time-scales in my field, it would be better for me to say *only* 3 million years, since in a geological sense this is a very short time-span. But I admit such processes are unimportant for people concerned about changes in our environment in the next few decades or even centuries.'

'However, the point remains', I continued, 'that technical, biological and geological processes cannot endanger atmospheric oxygen, at least as far as the next hundred or even thousand years are concerned. I therefore decided, with regard to my book, not to concentrate on oxygen but rather on carbon dioxide where the situation is far more critical.'

'If it's any comfort to you', Professor Schidlowski smilingly remarked, 'I had a similar experience. Originally I was a geologist. When I started working with atmospheric problems I also had the impression that oxygen was endangered. Our director, Professor Junge, who heads the Department of Air Chemistry, repeatedly told me that it was not the oxygen but carbon dioxide that was likely to cause problems. But it took months before I really understood that completely.'

'And then, everything seems so clear and obvious,' I remarked.

Professor Schidlowski nodded.

'But I did not tell you the clue to the story,' I continued. 'Just after I had decided on a new concept for my book, and started to concentrate on the carbon dioxide, I read the latest paper by Garrels, Mackenzie and Lerman, a copy of which I have seen lying on your desk.[47] If one follows the lines developed in this paper one realises that deforestation, which had been worrying me from the start, not only causes a dangerous increase in atmospheric carbon dioxide and a corresponding, though unimportant, short-term reduction in oxygen, but actually, through increased erosion, does have a long-term impact on atmospheric oxygen.'

'And to make these relationships understandable to your readers you are going to tell them a little bit about the history of the atmosphere?'

'Yes. That's the reason why I came here.'

2

The history of the atmosphere

General dimensions

The atmosphere surrounding our planet extends about 1000 kilometres above the surface. Its height corresponds to slightly less than one sixth of the Earth's radius, its volume is about half that of our planet. Its total mass, however, is negligibly small when compared to the Earth's mass. It is only about one millionth of that of our planet. The total mass of water on Earth, including that of frozen ice as in the glaciers etc., is about 250 times that of the atmosphere. These proportions 1 to 250 to 1 000 000 should be kept in mind when considering the history of our planet as outlined in the following paragraphs.

In early geological ages the masses of the atmosphere and the water were certainly smaller than at present. On the other hand the total mass of gases which were once part of the atmosphere and have been lost through rock-forming processes or have escaped to space was considerably larger than the present mass of the atmosphere.

Most of the air mass, i.e. 99.5 per cent, is concentrated within the lowest 40 kilometres of the atmosphere. The remaining 0.5 per cent is distributed among the 960 kilometres above. The lowest layer, where clouds form, and where the weather develops, is called the troposphere. It extends to a height of about 10 to 12 kilometres in mid-latitudes and up to 18 kilometres in the tropics. Nearly 90 per cent of the total air mass is in this layer.

The next layer above it is called the stratosphere, where only trace amounts of water vapour and virtually no clouds exist. In this height region, extending to about 50 kilometres above the Earth's surface, the ozone layer is found. The origin of this 'shield' which is vitally important for life on Earth will be discussed later in detail.

Different names are used for the atmospheric layer above the stratosphere, depending on the scientific point of view. The term ionosphere is commonly used, and it indicates that electrically charged (ionised) particles exist in this rarefied gas medium. Some of the ionospheric layers reflect electromagnetic waves and are therefore important for radio communication.

The density of the atmosphere decreases rapidly with height. Beyond the stratosphere the air pressure reaches values which, in a laboratory on the Earth's surface, would be considered a reasonable vacuum. There is no sharp boundary between the atmosphere and space, which is itself not entirely empty. Moreover, there is a gradual transition. At about 1000 kilometres altitude light gases such as hydrogen and helium can escape into space. On the other hand there is a permanent influx of matter, mainly protons (nuclei of hydrogen atoms) from the so-called solar wind, as well as material from meteorites. The quantitative relationship of gain and loss from these processes is not known precisely. It is generally accepted that since the Earth's formation, loss of matter has been prevailing.

The capability of a planet to build up a stable atmosphere depends on its mass and temperature. The greater the mass, the stronger its gravitational field and thus the greater the attractive forces that act on the gas particles. The higher the temperature, the more rapid is the movement of the gas particles and thus the higher the probability of their escaping into space. For the temperature conditions experienced by the inner planets in the solar system, the gravitational field of a planet such as the Earth is not sufficient to hold the lightest gases (hydrogen, helium), whereas Jupiter, being considerably larger in size and more distant from the Sun (colder), loses only minute amounts of hydrogen.

Small planets (planetoids) orbiting around the Sun, mainly between Mars and Jupiter, and meteorites have no atmosphere at all, because their gravitational fields are too weak. At best, a few traces of gaseous or liquid compounds might be held in pores of rocks in the interior of these objects. Eventually these remnants evaporate when such a body approaches the Sun and thus gets warmer.

Our planet is 4.6 thousand million years old. The idea that it originated in its entirety from a fiery-liquid sphere is virtually no longer accepted. During the last few decades scientists have come to the conclusion that our planet was formed from a conglomeration of small planetoids and meteorites. The circumstantial evidence for this was not provided by astronomers but mainly by geochemists and cosmochemists, and their most important piece of evidence is the composition of our atmosphere. The starting point for their argument is the fact that the composition of the Earth differs considerably from the well-known distribution of the chemical elements in the solar system and the universe. Our solar system consists of nearly 90 per cent of hydrogen, 9 per cent of the noble gas helium, and merely 1 per cent of all of the other ninety chemical elements combined. One can easily explain why the Earth contains much less hydrogen and helium and considerably more of the other elements. As was described in the previous section, its gravitational field is simply too weak to retain these two light gases.

Besides hydrogen and helium however, there are also other chemical elements, in particular the other noble gases, which are much rarer on Earth than elsewhere in the universe. If our planet originated in its entirety, the abundances of neon, crypton and xenon would have to be greater by a factor of millions to thousands of millions, as compared to those found today. These gases, unlike helium, are so heavy that only negligible quantities can have escaped to space. Since they are inert they are not lost through chemical processes. They are not radioactive and hence are not converted into other elements. Thus if they were present when the Earth was formed they should still be in the atmosphere.

In fact, only the noble gas argon is found in an appreciable quantity, but nearly exclusively in its isotopic form with the atomic weight of 40. This isotope however is not an original constituent of the atmosphere. It has been formed from the decay of radioactive potassium and has gradually accumulated in the atmosphere throughout geological time. Its present atmospheric concentration corresponds to that to be expected from the abundance of radioactive potassium on Earth. Hence argon-40 is not related to our circumstantial evidence concerning the gases present when the Earth was born.

Thus, disregarding argon, there are only traces of noble gases in the Earth's atmosphere, just as on a planetoid or a meteorite. The explanation, that our planet was formed mostly from the conglomeration of small planets and meteorites, stripped of most of their gaseous constituents including hydrogen, therefore appears plausible. Similarly one could argue that, through unknown cosmic processes, these elements had already been diluted in those regions of the cosmic gas cloud, from which, close to the Sun, Venus, Earth, Mars, planetoids and meteorites were formed. If either of these explanations holds true, our planet was most likely bare or nearly bare of a primary gas envelope when it was born. According to this interpretation our atmosphere has been forming since that time from gases emanating out of the Earth's interior.[2] STOP

The ancestors of our atmosphere

Celestial bodies contain certain quantities of radioactive elements. The energy produced by these elements provides efficient interior heating. In a small meteorite this heat rapidly reaches the surface by conduction and is radiated into space. A large body insulates more effectively. Thus its interior is much warmer than its surface. When the Earth was formed as a conglomerate of small bodies, a tremendous warming of the interior of the newly-formed planet must have resulted from these processes and from gravitational effects. Many chemical reactions were thus initiated, some of whose products were gaseous. First of all water, chemically linked as water of crystallisation, was liberated through this heating process. This water vapour and other gaseous products have been emanating from the earth's interior since then through volcanic eruptions and outgassing. They are the raw material from which the gas and water envelope of our planet has been formed.

Since radioactivity gradually decreases, one can assume that in earlier geological times the Earth's interior was hotter than it is at present. Thus volcanic eruptions and outgassing are likely to have been more frequent and stronger than nowadays. One can therefore assume that the primeval atmosphere formed relatively quickly, soon after the planet was formed. Its composition was that of the volcanic gases which in their part were determined by the nature of the molten rocks in the Earth's interior.

The present composition of volcanic gases is well known and compiled in the following table. There is no evidence of any essential change in this composition during the course of the past three to four thousand million years. Several scientists maintain, however, that very early in the Earth's history it might have been different. Nowadays volcanic gases certainly contain no free oxygen, but they are mainly fully oxidised compounds such as water vapour or carbon dioxide. If during the early times the molten rocks near the Earth's surface contained more free iron or other substances with a high affinity for oxygen, the outgassing products might have had a smaller fraction of fully oxidised compounds. On the other hand, *more* free hydrogen and hydrogen compounds with carbon (methane), sulphur (hydrogen sulphide), and nitrogen (ammonia) are likely to have emanated. Such an environment, richer in hydrogen and deficient of oxygen, is classified by scientists as a 'reducing' environment. The term 'more' used here with regard to hydrogen and its products must be understood in a relative sense. The atmospheric content of methane and hydrogen is measured in parts per million today (see table 1 on page 171). If in the primeval atmosphere the abundance of reducing substances was only about

Average composition of volcanic gases[3]

Main component	Chemically related admixtures	Percentage
Water vapour (H_2O)	Hydrogen (H_2)	80
Carbon dioxide (CO_2)	Carbon monoxide (CO), methane (CH_4)	12
Sulphur dioxide (SO_2)	Sulphur (S_2), hydrogen sulphide (H_2S)	7
Nitrogen (N_2)	Ammonia (NH_3)	1

Furthermore, small quantities of hydrochloric acid (HCl), hydrofluoric acid (HF), argon (A), and helium (He) are also present, the last being produced by radioactive decay of uranium and thorium. Unlike argon, the light helium does not stay to accumulate in the atmosphere, but diffuses upwards and eventually escapes into space.

the same as that of carbon dioxide in the present atmosphere, i.e. several hundredths of a per cent, this could have been of great importance for several geochemical processes and indeed for the origin of life itself.

In fact, such a slightly reducing atmosphere may probably have existed for a 'short' time only, say some hundred million years. Free hydrogen and hydrogen liberated from hydrogen compounds by the Sun's ultraviolet radiation (so-called photolysis) escaped to space. Outgassing of hydrogen and hydrogen compounds decreased as the composition of the molten rocks near the surface and hence of the volcanic gases gradually approached their present conditions. Thus already 3.5 to 4 thousand million years ago a 'neutral' (neither reducing nor oxidising) atmosphere is likely to have evolved with water vapour, carbon dioxide, nitrogen, and argon being the main constituents.[4]

The largest fraction of the water vapour condensed, gradually forming the envelope of liquid water (the hydrosphere) which is unique in the solar system. Our 'blue planet' owes this nickname to being exactly at the right distance from the Sun. Thereby, our environmental conditions are fundamentally different from those on our neighbouring planets. The surface of Venus, which is closer to the Sun, is too hot for the formation of liquid water, whereas Mars, at a greater distance, is too cold for liquid water to exist.

With the formation of the hydrosphere, the Earth possessed an ideal solvent where atmospheric constituents could react with minerals in the crust. In particular, most of the carbon dioxide, outgassing from the Earth's interior, was dissolved in the oceans thereby forming carbonates such as limestone. As the American Nobel Laureate Harold C. Urey has shown, this process is of fundamental importance for the composition of our atmosphere. Since there are huge amounts of liquid water on Earth, carbon dioxide can react with the minerals of the crust. Thus on a long time-scale, *atmospheric carbon dioxide cannot accumulate to very much higher levels than those found at present.*[5] However, processes such as gas exchange between atmosphere and ocean, ocean mixing, and reactions with the crust material do require a considerable amount of time. For time-scales short when compared to geological times, say for hundreds to thousands of years, an increased carbon dioxide influx would certainly cause the atmospheric carbon dioxide concentration to rise.

Among the other gases emanating from the Earth's interior, sulphur dioxide, and hydrochloric and hydrofluoric acids were almost completely dissolved in the oceans. Thus both soluble (seasalt) and nonsoluble minerals were formed. Nitrogen, although only a minor fraction of the volcanic gases, gradually accumulated in the atmosphere. Due to its low reactivity under existing conditions it was not dissolved in the oceans and thus became the main constituent of the atmosphere. The noble gas argon, outgassing in considerably lesser amounts, accumulated too. Oxygen, however, the second main constituent of our air, certainly did not enter the atmosphere in this way.

'Air' without oxygen

Many details of the early history of our atmosphere, a rough picture of which was given in the previous section, still remain controversial and unclear. About one essential point, however, there is considerable agreement. *Until about two thousand million years ago the atmosphere of our planet did not contain any appreciable amounts of free oxygen.* Scientists from different fields, through independent research efforts, have come to this conclusion.

Above all, it was the geologists who furnished evidence for this statement. Old sediments, which have not been exposed to the atmosphere for the last two thousand million years, contain large quantities of minerals that are not stable under present atmospheric conditions. In particular pyrite (FeS_2), a brass-like yellow crystal, decays within a few years, forming ferric oxide and sulphur dioxide, when exposed to air. The famous Witwatersrand gold deposits in South Africa consist of many horizontal layers of solidified gravels and sands, of up to 6 kilometres in depth. Besides the main component quartz these sands also contain many grains of pyrite. The crystals found in the deposits appear to have been rounded by wind and water, and thus most likely had been exposed to the atmosphere for some time before being deposited two to three thousand million years ago. They do not however show any indication of oxidation processes. Grains of gold and pitchblende are found at the bottom of the sediment layers since they are heavy. Pitchblende is a mineral that is also oxidised when exposed to the atmosphere. Similar deposits to those of Witwatersrand are also found in Canada and Brazil. In view of the widespread distribution of such formations, the thickness of the deposits, and the fact that these also contain other oxidisable minerals, the possibility of accidental localised conditions can certainly be excluded. This led the doyen of German mineralogy, P. Ramdohr, to conclude as early as 1958, in spite of doubts of other geologists, that the Earth's atmosphere did not contain oxygen when these sediments were formed.[6]

On the other hand, unambiguous evidence for the presence of oxygen in the atmosphere is found only in sediments younger than about 1.8 thousand million years. Here in particular the 'red beds' (sandstones and shales) are to be mentioned, the rusty colour of which is due to admixtures of ferric oxide. There is no fully oxidised 'red' iron found in older continental sediments.[7]

While in 1958 the statement that two thousand million years ago the Earth's atmosphere was devoid of oxygen was considered by several geologists to be a heretical hypothesis, it was not at all surprising to scientists working in other fields. In particular, those involved in research into planetary evolution and the origin of life found their assumptions and theories confirmed. Some of these theories had in fact been published much earlier. It would indeed be hard to explain why an oxidising atmosphere should have existed from the very beginning on Earth whereas our neighbouring planets possess hardly any free oxygen at all. Moreover, if oxygen were not consistently replenished it would rapidly disappear from the atmosphere. Being very reactive it reacts with the minerals contained in the crust. Oxygen certainly could not have escaped from the Earth's interior because of its rapid interaction with other chemical elements. Consequently the question arises, where has the free oxygen come from?

As long as no green plants existed only one process could have been responsible for the production of oxygen, namely the dissociation of water vapour through ultraviolet radiation. This process, called photolysis, occurs at great heights where atmospheric densities are sufficiently low for part of the liberated hydrogen atoms to escape into space before colliding with oxygen atoms and thus reforming water molecules. The remaining oxygen atoms, which, put simply, are unable to find a partner, accumulate in the atmosphere. The American scientist H. C. Urey however, whose thoughts on planetary evolution we have followed in the previous chapter, shows that this oxygen production through photolysis is a self-limitating process. Free oxygen is opaque to ultraviolet radiation at exactly those wavelengths necessary for the dissociation of water vapour. Thus if a certain amount of oxygen is formed in the high atmosphere the water vapour underneath it is protected from further photodissociation. Whether or not oxygen produced through photolysis could accumulate in the near-surface layers at all is a subject of some controversy. If the answer is yes then it must surely have been only in trace amounts. L. V. Berkner and L. C. Marshall, assessing the consequences of this 'Urey-effect', found that at most one thousandth of the oxygen presently found in the atmosphere could have originated from photolysis.[8] On the contrary, the American physicist R. T. Brinkmann is of opinion that up to 25 per cent of the present atmospheric oxygen might have been produced through photolysis. He admits, however, that this hypothesis conflicts with well-known geological and biological facts, and speculates that the oxygen liberated through photolysis undergoes chemical reactions in the high atmosphere (in particular with hydrogen carried along with the 'solar wind'), without reaching near-surface layers.[9] The fact that nearly no oxygen is found in the atmospheres of Venus and Mars is a further argument against the proposition that photolysis is the dominant mechanism for producing this chemical element in the Earth's atmosphere.

Biologists and biochemists finally came to the conclusion that life must have originated in an environment devoid of free oxygen. The synthesis of increasingly more complex chemical substances would have been impossible otherwise, for even the most simple building blocks of a living substance would have met an untimely end by being 'burnt', that is, oxidised. In fact, the metabolism of the most primitive organisms also indicates that these organisms must have originated when no oxygen was present. These anaerobic (oxygen-avoiding) organisms cannot thrive under present atmospheric conditions. They meet their energy requirements through fermentation. Moreover only an atmosphere devoid of oxygen could provide the 'radiative climate' which was necessary for the origin of life.[10] But to understand this we first must deal with the relationships between atmospheric composition and the radiation penetrating to the Earth's surface.

Radiation 'currencies' and atmospheric 'customs duty'

The Sun is by far the most important source of energy for our planet. Its radiation spectrum is very broad. It extends from X-rays, with wavelengths of the order of a millionth of a millimetre (i.e. a nanometre, 10^{-9} m), through ultraviolet and visible light, infrared radiation, microwaves to radio emission, with wavelengths up to several kilometres. For every radiating body the energy distribution of its radiation spectrum is determined by its temperature. The warmer the radiator, the more the region of bulk radiation is shifted toward shorter wavelengths. The Sun, with a surface temperature of about 5500 degrees centigrade, radiates its maximum flux as visible light, with decreasing amounts of radiation towards both ends of the spectrum.

The radiated energy is transmitted in small nondivisible units of energy that are given the name 'quanta'. For every wavelength, each quantum has a well-determined value which decreases with increasing wavelength. In a metaphorical sense one could say that every wavelength can merely pay in its own individually assigned 'energy-currency'. Thus the shortwave X-rays could deal in pounds sterling, ultraviolet radiation in dollars, visible light in German marks, infrared radiation in Austrian schillings, and radio emission in Italian lira. In this sense the Sun spends most of its money as marks and then, in descending order, dollars, schillings, pounds and lira. Of course this is only a very approximate picture. In fact, not only every spectral region, such as visible light, but every single wavelength has its own currency.

To initiate physical or chemical processes not only is a certain amount of energy required but also this energy must often be paid for in a well-determined currency. In other words, it is not only important that solar radiation (and hence energy) penetrates to the Earth's surface, but also what spectral region this radiation is associated with.

All atmospheric constituents differ in their transparency to radiation of various wavelengths. One could say that they demand differing customs duties depending on which currency the transit trade is based. Carbon dioxide for instance lets marks and dollars (visible and ultraviolet light) pass nearly duty-free, whereas it demands considerable dues for schillings (infrared radiation). Although the present abundance of atmospheric carbon dioxide is only about one thirtieth of a per cent this has a noticeable effect. The hot Sun radiates, as has already been mentioned, most of its energy as visible light, which can pass through the various gases that constitute the atmosphere almost unattenuated. The much colder Earth, however, emits mostly radiation of longer wavelengths, that is in the infrared, which are partially absorbed by the carbon dioxide. Thus without carbon dioxide in the atmosphere the Earth's night surface would cool rapidly. Changes in the atmospheric carbon dioxide content are thus likely to cause feedback effects in the radiation and energy budget of our planet.

Not only carbon dioxide but also the other constituents of the atmosphere are more or less transparent to visible radiation. Only water vapour and in particular condensed water (clouds, fog) and atmospheric dust, a consequence of storms, volcanic eruptions as well as of human activities, demand certain customs dues in this spectral region. Water vapour also absorbs strongly in the infrared, which is the reason for a stronger radiative cooling during clear nights as compared to cloudy conditions.

X-rays and shortwave ultraviolet radiation with wavelengths less than 100 nanometres get absorbed in the thin gas layers of the upper atmosphere and thus do not penetrate below about 50 kilometres height. This radiation is so energetic that it can detach single electrons from the electron shells of atmospheric gas atoms. Thereby, as the radiation is absorbed, the atoms are ionised (electrically charged). Thus the outer layer of the atmosphere where these processes occur is called the ionosphere.

Ultraviolet radiation between 100 and 200 nanometres wavelength is mostly absorbed by oxygen molecules in the upper layers of the stratosphere. In this process oxygen molecules are split into two single atoms, most of which recombine again to form oxygen molecules, particularly during the night in the absence of solar radiation. Some react with oxygen molecules (O_2) however, forming ozone (O_3), the triatomic form of oxygen. Ozone for its part strongly absorbs ultraviolet radiation of wavelengths between 200 and 300 nanometres thereby becoming decomposed into regular diatomic molecules and single atoms of oxygen. Moreover, oxygen atoms can react with ozone forming two oxygen molecules ($O + O_3 = 2O_2$). Furthermore, ozone is decomposed through chemical reactions with various atmospheric trace gases. Through all these processes which are linked together in a complex chemical system, ozone is continuously produced as well as decomposed. In the resulting equilibrium, about 20 to 50 millionths of the oxygen present in the stratosphere exist as ozone. This minute amount, however, is sufficient to absorb the total ultraviolet radiation from the Sun between 200 and nearly 300 nanometres wavelength.

This absorption of radiant energy consequently leads to an efficient heating of the stratosphere. Whereas the temperature of the troposphere decreases with height at a rate of about 6 degrees per kilometre and reaches a temperature of about -60 degrees centigrade at the lower boundary of the stratosphere, it increases again in the stratosphere. At 50 kilometres altitude, the upper boundary of the stratosphere, the temperature is about zero degrees which is nearly as warm as at the Earth's surface. This temperature distribution is the reason for the tremendous vertical stability of the stratosphere, where convection and cloud formation virtually do not exist. The gas exchange between the stratosphere and the underlying tropospheric layers is so weak that the average residence time of dust particles injected into the stratosphere (through strong volcanic eruptions for instance) is much longer than in the troposphere where vertical circulation and precipitation provide efficient scavenging.

The atmospheric ozone layer protects living beings at the Earth's surface from a radiation which otherwise would be lethal. Ultraviolet light of wavelengths between 260 and 290 nanometres is capable of decomposing the most important constituents of living substances, namely protein and nucleic acids. Certainly this dangerous radiation is unable to penetrate deeply into a multicellular organism, as plating or a thick fur do indeed provide a certain amount of protection. For human skin however, a strong dose causes cancer, and plant leaves can also be damaged. The micro-

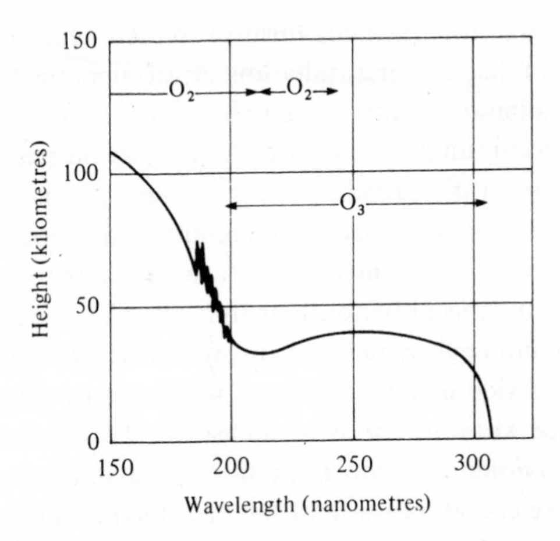

Figure 6(*a*) Penetration depth of solar radiation into the Earth's atmosphere. The curve shows the height at which solar radiation is reduced to about one third of its extraterrestric intensity (exactly: to $1/e = 36.8$ per cent; e stands for the base of the so-called natural logarithm.). Solar radiation with wavelengths less than 295 nanometres does not reach the Earth's surface. Radiation with wavelengths less than 200 nanometres is mostly absorbed by oxygen molecules (O_2), whereas that with wavelengths between 200 and 300 nanometres is predominantly attenuated by ozone (O_3).

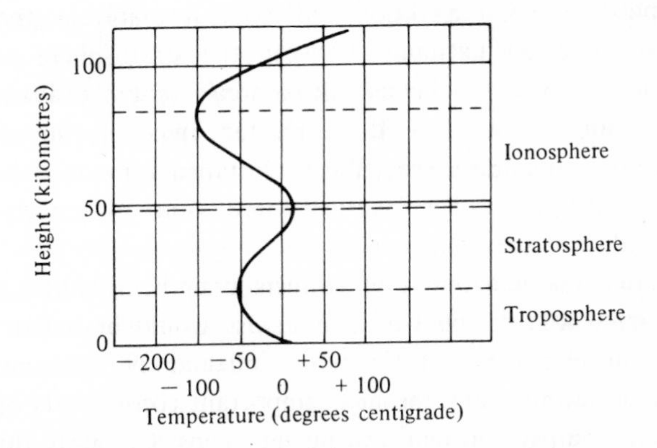

Figure 6(*b*) The temperature profile of the atmosphere. In the troposphere the temperature decreases with increasing height. In the stratosphere it increases up to a temperature maximum around fifty kilometres, above which temperatures decrease again. Above eighty kilometres, a tremendous temperature increase is observed. Due to the low gas densities at these heights, however, 'temperature' has another meaning than the one we are used to.

scopically small unicellular organisms which came into being as the first forms of life are naturally the most imperiled. For these, ultraviolet radiation, severely disturbing their metabolic processes and cell division, finally means death. This is the reason why this form of radiation is commonly used for sterilisation purposes.

Thus for as long as no free oxygen, and hence ozone, existed in the atmosphere of our planet, it was not only the chemical conditions but also the radiation climate that differed fundamentally with our present environment. No life was possible on land surfaces then. Only under a few metres of water would life have been possible since water demands high customs dues in dollars for the harmful ultraviolet radiation. On the other hand, however, it was in particular the energy of these ultraviolet-dollars that provided the important capital stock for a variety of chemical reactions necessary for the origin of life.

Life by chance?

The fact that the Earth is orbiting the Sun at exactly the right distance for liquid water to exist, is the most important accident for the origin of life. The question as to whether or not life might be possible in an environment based on completely different biochemical conditions, with say liquid ammonia as a solvent, remains a topic for mere speculation.

It was the Soviet biochemist Alexander I. Oparin and the British biologist and geneticist J. B. S. Haldane who, during the time between the First and Second World Wars, put forward a theory on how terrestrial life might have developed in the presence of water and a reducing (or at least non-oxidising) atmosphere.[11] In 1953 this theory was experimentally confirmed by S. L. Miller, a young American who then held a scholarship for research with H. C. Urey. For one week Miller had a mixture of methane, water vapour, ammonia and hydrogen, likely constituents of the reducing primeval atmosphere, circulating in a sealed glass-bulb permanently exposed to electrical discharges. The analysis of the substances formed in this experiment yielded sensational results. Miller found important building blocks for life, in particular amino acids.[12]

In the course of the following years Urey and Miller, and a large number of other scientists as well, repeated these experiments under varying experimental conditions. Different compositions of the 'atmosphere' and various energy sources have been tested. In addition to electrical discharges simulating thunderstorms, ultraviolet radiation, which was likely to have provided the largest fraction of energy in the primeval atmosphere, and radioactivity, to simulate the radioactive material contained in the Earth's crust, have been tried out.[13] The results of all these experiments showed undoubtedly that under the conditions likely to have existed several thousand million years ago, all the important components of organic substance could have been formed in this way.

These building blocks dissolved in water reacted with each other. There, protected from destruction by ultraviolet radiation, macromolecules, consisting of hundreds or thousands of atoms, could form. Minerals such as clay, lava or siliceous earth might have acted as carrier substances and ion exchangers. Finally, through accumulation of these macro-molecules in the water, an agent arose which scientists commonly call 'primordial soup'. In favourable conditions, such as in shallow inland waters or secluded lagoons with weak water exchange, a rich broth is likely to have formed. This must have had an almost indefinite lifetime due to the absence of decomposing microorganisms and free oxygen that would oxidise such dissolved substances. The primordial soup apparently must have been rich in nutrients (carbohydrates) as well as structural components such as proteins and nucleic acids (the substances that carry genes). It is certainly imaginable that in such a solution, through randomly initiated reactions, all the enzymes and the other compounds necessary for the origin of primitive life could form. A solution containing all essential constituents of a living being, however, is a long way from a living being itself.

If we were to dismantle a thousand cameras, all of the same type, and then put all the components into a large box and shake it, there is not the slightest chance that one single camera will reassemble itself. Even if we were to shake the box for a hundred years, chance will not help us. The components of the microscopically small sub-units forming the vital parts of living cells however behave quite differently. As a matter of fact, they recombine 'by themselves', as numerous experiments carried out during the last two decades have shown. Even complete viruses, decomposed into their components, are capable of recombining to function normally without any external assistance.

These processes are in no way mysterious. They can be explained by well-known natural laws based on the chemical properties and the three-dimensional structure (the so-called stereochemical properties) of proteins and nucleic acids.[14] Thus, one can further imagine, that in the beginning such single components, or more exactly compounds that were to constitute components in the future, originally combined by accident according to their physical and chemical properties, until finally a 'system' was established capable of reproducing itself (and increasing its numbers) and maintaining itself within its environment. By present standards, this system must have been a very primitive one, but it was a 'living being'.

With this as an excuse, we have entered the fascinating field of the self-organisation of the biological macro-molecules, which owes its major impulse to the theories put forward by the German Nobel Laureate Manfred Eigen. To discuss these here in detail would be outside of the scope of this book. For Eigen, the origin of life and hence the development from macro-molecules to microorganisms is only 'one single step among many others; elementary particles to atoms, atoms to molecules, and similarly unicellular organisms to multicellular ones and finally the central nervous system of man. Why should we regard this very step from macro-molecules to the unicellular organisms with greater respect than any other?' he concluded.[15]

Assessments of the probability that life originated in this way by chance, i.e. by accidental combination of more or less 'fitting' components, vary widely. According to the French Nobel Laureate Jaques Monod, the probability is so small that the crucial event is likely to have occurred only once. 'This would mean that prior to the event, the probability of its occurrence was virtually zero.'[16]

This opinion is not generally accepted, however. R. W. Kaplan, a German biologist, arrived at considerably higher probabilities for the coincidence of conditions favourable for the origin of life. Assuming that the first living beings thus formed were less complex than the most primitive of the organisms that exist today, he started calculating and came to the conclusion – surprising to himself, as he states – that a lake of no more than ten square kilometres of primordial soup would be necessary to provide a good chance for the origin of life.[17]

According to this view it is assumed that during the early history of the Earth the crucial event did not only occur once but many times. Thus, while the atmosphere was still in its reducing state, different species of living beings, with different biochemical systems, might independently have evolved.[18] Finally, as the result of competition for living space, food, etc., one species (or a group of closely related species) is likely to have dominated, since all present living beings are biochemically related to each other, signifying a common origin.

From fossil food to photosynthesis

The first living beings were born into a land of milk and honey. The primordial soup provided plenty of food, and there was no deadly competition with other species which presumably did not exist at the beginning. Only that way could the still imperfect creatures maintain and gradually harmonise their different systems and adapt to their environment.

Such 'adaptation' is *not* a process regulated by the living beings themselves. It happens rather through small errors, so-called mutations, occurring in gene transfer. For a particular being the occurrence of a mutation is a matter of chance. For a large number of individuals, however, statistical laws hold, and probabilities for the occurrence of a certain mutation or for the coincidence of mutations can be calculated reliably.

To species already well adapted to their environment, a mutation generally proves a handicap, a 'congenital defect'. To the first organisms however, which due to their poorly matched systems could hardly survive even in a land of milk and honey, mutations might often have proved advantageous. Thus, either a better adaptation to the environment was achieved or some sort of interior adaption, a better coordination of their components, that had initially joined together at random. Species subject to such advantageous mutations were more viable than their nonmutated relatives. Thus they could better surmount the process of selection which finally led to the survival of only those species which were best adapted to their environment.

The first primitive organisms lived on chemical compounds dissolved in the primordial soup, which had been produced long before life originated and continued to be produced in limited quantities. These substances, containing the stored energy of ultraviolet solar radiation which had led to their formation, could be used up by the primitive organisms to meet their energy requirements in the same way as we use fossil fuel, such as coal or oil, containing solar energy stored from ages long past. Thus these organisms lived, metaphorically speaking, on fossil food.

Initially the energy contained in the nutrients was liberated through fermentation processes – and many primitive microorganisms are still solely dependent on this kind of energy production today. But even higher respiring organisms did not forget in general about fermentation. They fall back upon it as an auxiliary process, for instance when the oxygen supply is not sufficient as in a muscle that is heavily stressed for a long time.

One can assume that the primitive organisms gradually became better adapted to their environment and thus multiplied rapidly. But through their own metabolisms *they now began to alter their environment.* Where favourable conditions existed, the demand for nutrients soon became greater than its rate of formation from solar energy, and the stock of fossil food was gradually exhausted. Famine was lurking.

Certainly this was not the case everywhere and all at once, otherwise fermenting microorganisms would probably not have survived up to the present day. But running out of nutrients implied either the necessity for another adaptation process to account for the new environmental conditions created by the organisms themselves, or the likelihood of perishing.

The adaptation to a changing environment was again governed by the random game of mutations and selection. A congenital defect in the metabolic system under the old conditions of ample supply of nutrients could well mean in the new situation with the likelihood of famine the ability of just surviving by thriving on a new substance which formerly could not be utilised by the 'normal' species. And again, this key mutation just enabled mere survival and had to be followed by additional mutations until a reasonable coordination of the changed single components, and thus the adaptation to utilising the new nutrient, was again achieved.

In this manner different methods of fermentation and also other processes of energy production must gradually have evolved, not all of which may have survived. But sooner or later all the fossil food must have been exhausted at certain places before the utilisation of a different kind of energy producing reaction was achieved. In many cases this was certainly the end and wiped out all life in this particular region, until once again a 'fundamental invention' was made.

Today we human beings, facing a situation of fossil fuel – hence stored solar energy – becoming short and thus more expensive, are trying to switch over to other kinds of energy. Among others, solar energy is a likely alternative. About four thousand million years ago our unicellular ancestors solved their energy crisis exactly the same way. When they ran out of fossil food they switched over to directly utilising incident solar radiation.

Ultraviolet radiation, which had played such an important role in the formation of the primordial soup, was unsuitable for this purpose because it would destroy the living beings. Rather, visible light, with longer wavelengths capable of penetrating water to depths inaccessible to ultraviolet radiation, was the solution. By means of suitable pigments its energy can be absorbed and utilised for liberating hydrogen from various compounds, which subsequently acts as a source of energy for biochemical processes.

Since solar energy is available only during daylight hours, a storage mechanism is necessary to meet the nightly requirements. This is achieved by chemical reactions between the liberated hydrogen and carbon dioxide. Carbohydrates (glucose, starch) thus produced can be decomposed (fermented or 'burnt') with a gain of energy whenever needed. A precondition for this kind of energy production and storage, so-called photosynthesis, is the existence of sufficient carbon dioxide dissolved in the water.

It is assumed that photosynthesis was 'invented' by ancestors of the present purple bacteria which produce their hydrogen through decomposition of hydrogen sulphide; besides carbohydrates the end-products of this process are sulphur or sulphates and water. All plants producing hydrogen through decomposition of water, with oxygen being exuded as a waste product, are likely to originate from ancestors of this type. The transition to utilising this important reaction, which requires considerably more light-energy than the decomposition of hydrogen sulphide, enabled organisms maintaining photosynthesis to populate all environments habitable at that time.

The oldest organisms classifiable as plants apparently resembled the present blue-green algae. Since they originated at a time when no free oxygen existed in the atmosphere they had to decompose their carbohydrates through fermentation rather than respiration. There are still varieties of blue-green algae which thrive in the absence of oxygen (under anaerobic conditions).[19] Blue-green algae belong to the large group of primitive unicellular beings that do not possess a cell-nucleus – the so-called procaryotes. In the ocean they form carpet-like layers gradually being calcified. In South Rhodesia such deposits of limestone, so-called stromatolithes, were found which are nearly three thousand million years old. Thus the evolution of this oldest family of plants obviously dates back to a very remote past.[20]

How old is life?

The fact that terrestrial life is much older than three thousand million years has essentially been proved only since the mid-sixties. Originally many scientists had doubts, since multicellular organisms have only existed some seven hundred million years, and the whole impressive evolution from microorganisms to man has been accomplished during this period. It was hard to understand that before this time unicellular beings should have already existed five times that long without any exciting evolution having taken place – and that on the other hand the origin of life from lifeless matter should have occurred within less than one thousand million years. To many biologists, such a time schedule of evolution appeared incredible.

The above-mentioned Rhodesian stromatolithe limestones were

found as early as 1941, but at first their age of nearly three thousand million years was doubted. During the sixties however, more evidence was presented that life dates back to much earlier times than previously assumed. And last but not least one began to comprehend *why* it took so long for multicellular organisms to evolve: that this evolution was possible only after the organisms themselves had provided the proper conditions by altering the atmospheric environment.

The fundamental work of the Americans E. S. Barghoorn and J. W. Schopf provided important evidence. In the mid-fifties they systematically searched for traces of microscopically small primeval creatures. Applying the most modern scientific equipment and research techniques, with the assistance of a huge staff that had been trained in different fields of science, their research efforts were quite successful.

To begin with this team concentrated on Canadian sediments, a little less than two thousand million years old. In these layers the electron microscope revealed a large number of features which apparently were petrified remnants of bacteria and blue-green algae. In many cases it could be proved unambiguously that these organisms could not have invaded these layers after the sediments were formed.

The cogency of such findings alone might not have been sufficient to convince sceptical scientists that these were really fossils, that is remnants of primeval organisms and not some sort of random array of inorganic origin. Such features often at first sight appear to look like primitive unicellular creatures. Thus the American team searched for supporting evidence to prove the biological origin of their findings.

One of their methods was a statistical investigation of the size distribution of the features in question. As far as bacteria of one particular species are concerned – and in general a large number of petrified bacteria rather than just a single one can be expected to be found – they should all be about equal in size and shape. Hence the size distribution reveals a pronounced narrow maximum whereas the sizes of inorganic features such as blobs from petrified bubble formation show a considerably wider range of scatter.

A further method is to search for substances that can be classified as chemical fossils, i.e. remains of decomposition products of organic substances, among these chlorophyll, the pigment necessary for photosynthesis. Recently however it turned out that chemical fossils soluble in water are of not much use. For, as J. W. Schopf pointed out in a very critical review of his own work, 'it is extremely difficult, if not impossible, to establish that these compounds date from the time of sedimentation rather than having been introduced (e.g. via ground waters) in the relatively recent geological past'.[21] As for carbon compounds insoluble in water, one generally can assume that they have been part of the sediments in which they are found from the very beginning. In most cases however, it is difficult to prove that these are decomposition products of biological matter rather than substances of inorganic origin.

A further piece of evidence for the biological origin of carbon compounds embedded in the Canadian sediments was found using mass spectrometry. Carbon which has become part of the organic substance through photosynthesis can be distinguished from other carbon by its isotopic composition. It is 'labelled', as it were. Carbon dioxide molecules formed with the carbon isotope of atomic weight 12 are more mobile than those containing heavier carbon-13. Hence they penetrate cell membranes more easily and are preferentially taken up by photosynthesis. Consequently, living beings and their decomposition products contain relatively less carbon-13 than exists as atmospheric carbon dioxide or in inorganic carbon compounds. Thus the measurement of the isotopic composition can provide important information as to whether or not a certain carbon sample is of biological origin.

Along these lines Barghoorn, Schopf and their colleagues put together a series of independent analytical methods that supplemented each other. Each method taken by itself might not be considered conclusive, but the consistent results from all of these provided a fairly certain evidence as to whether the old sedimentary rocks contained microfossils. Having proved this for the Canadian sediments dating back about two thousand million years,[22] Barghoorn and Schopf started analysing much older sediments from South Africa.

When this programme started in the early sixties Barghoorn and

Schopf were not at all sure that they would find any remnants of organisms in such old rocks since at that time, nobody knew how long life dates back to on Earth. In order to find out, they wanted to trace it back as far as possible, even to its origin if feasible.

At the time when the American scientists began their research work on African sediments, a young German geologist was also active in South Africa, unaware that he too would soon pursue the same question. He was Manfred Schidlowski, already introduced as head of the research group working on the history of the atmosphere at the Max-Planck-Institut für Chemie in Mainz. 'I was working for the Witwatersrand gold and uranium mines', he told me, 'when I read Ramdohr's famous article which proved, just on the evidence of the geology of the Witwatersrand deposits, that two thousand million years ago our atmosphere must still have been deficient of oxygen. [See page 32 of this book.] This was my first aquaintance with a field which later was to become the subject of my scientific work. I was fascinated by Ramdohr's conclusions, and since I was working for Witwatersrand mines I wrote to him asking whether he recommended more research be done there. His answer was very encouraging. Thus I started, at first just as a hobby, collecting rock samples which I brought back home in 1963 after having finished my job in South Africa.' With this kind of luggage Schidlowski had a good ticket for his scientific career. He joined Ramdohr's team at Heidelberg University to analyse his rock samples under the guidance of this distinguished mineralogist and petrologist.

'We were mainly interested in the so-called thucholithe then, a carboniferous substance found in the Witwatersrand deposits,' he continued. 'This mineral was obviously formed from originally liquid or gaseous hydrocarbons polymerised, baked together so to speak, by radiation from uranium. To a geologist, a mere inspection of the deposits makes it clear that these hydrocarbons have been part of these sediments since their formation and that they cannot have been introduced at a later time. Since these deposits are a lot older than two thousand million years, and since hydrocarbons are usually produced from decaying organic substances, we had to ask ourselves: Was there already life on Earth some two thousand million years ago with so many living beings existing that considerable amounts of organic carbon could accumulate in the sediments? In 1963, when I returned with my rock samples, this was still an open question. In order to prove that the carbon found in the thucholithe is indeed of biological origin we had to look for evidence.'

In close collaboration with scientists from other universities, Schidlowski used the methods applied by Barghoorn and Schopf which he developed further. By means of an electron microscope he could find clearly evident structures in his rock samples, most likely remnants of unicellular organisms or bacterial spores. Together with J. Hoefs of Göttingen University he proved that the abundance of the isotope carbon-13 was reduced in his samples, an indication for the biological origin of the carbon. Most impressive, however, were the results of the chemical analyses carried out with A. Prashnowsky of Würzburg University. By means of paper chromatography, a technique very suitable for detecting extremely small amounts of admixtures, they discovered a total of seventeen compounds of evidently biological origin in the carboniferous substance taken from the Witwatersrand deposits, among them amino acids and carbohydrates.[23]

'As far as I know', said Schidlowski, 'this was the first time that the presence of amino acids in such old rocks was proved.'

Meanwhile, Barghoorn and Schopf had also published the results of their studies on sediments from Transvaal and Swaziland which were up to 3.4 thousand million years old.[24] The American scientists did not succeed in tracing life back *to its origins*, still less in finding sediments *older* than life. In all the geological formations they analysed, microfossils could be found. In 1977, Barghoorn and Knoll detected even traces of bacterial cell division in sediments from Swaziland that were some 3.4 thousand million years old.[25] The question of when the *first* living beings evolved on our planet remained unanswered. But if Barghoorn, Schopf and their numerous collaborators did not completely misinterpret all their findings, it surely must have been more than 3.4 thousand million years ago.

Team work

Several of Schidlowski's papers appeared in journals of general interest read by scientists from different fields.[26] Thus Christian E.

Junge, who at the time had just become director of the Max-Planck-Institut für Chemie at Mainz, became aquainted with them. Junge was originally a meteorologist specialising in problems of air chemistry, and it is due to his work that air chemistry has become a well-established field of research. At that time Junge was about to bring together scientists from different fields to tackle problems related to the origin and evolution of the atmosphere. Of course the question of when the source of the atmosphere's oxygen, photosynthesis, started was of great importance. Schidlowski's scientific work dealt exactly with this question. Thus Junge invited him to join his team.

'Up to this time many scientists had somehow only just touched on the problems of atmospheric evolution', Schidlowski relates, 'and Junge's essentially new idea was to form a team that would be exclusively devoted to these questions. This task was interesting and attractive to me. Thus I came here to Mainz.' Together with the meteorologist and air chemist Junge and the geochemist Schidlowski, two chemists completed the team: Rudolf Eichmann, a specialist in mass spectroscopy, a method for measuring the isotopic composition of substances, and Jürgen Hahn, interested in the nitrogen cycle and a specialist in gas chromatography, a technique for measuring trace amounts of admixtures.

'Do you also have a biologist in your team?' I asked.

'No, but at times I think we should have one.'

Working in such an interdisciplinary team is very exciting, Schidlowski continued. Thanks to their differing backgrounds, the group members can supplement the work of each other in an ideal way and thus cover a tremendous field of knowledge. However compared to working alone, one is much more fixed to one's very special field.

'Sometimes I would also like to run the mass spectrometer, play with all the buttons and analyse my rock samples all by myself,' Schidlowski said, almost in a sad way. 'But then Eichmann always told me in a friendly but firm way to go back to my office. "With our technicians I can finish the measurements within a few weeks," he would say; and with a cunning smile he would add: "In order to get the results ready for publication as soon as possible one of us" (and of course he thereby meant me) "ought to retreat to develop models for a plausible interpretation of the data." After all such division of labour is extremely effective. To handle a complicated instrument such as a mass spectrometer a mere studying of the instruction manual is not sufficient. For the inexperienced, many months of effort are necessary to acquire the essential knowledge for a reasonable handling of the instrument. Thanks to our system of division of labour we could certainly advance much faster than any individual scientist.'

It is of course the geochemist who interprets the results of the analysis, and there is quite a lot to be interpreted. 'Our team has analysed the isotopic composition of carbon in rock samples of up to 3.3 thousand million years old,' Schidlowski continued. 'These were the oldest sediments known at that time. Our results have shown repeatedly that one fifth of the carbon contained in these sediments originates from biological sources. The same ratio is also found in much younger sediments. Thus we came to the conclusion, surprising as it was then, that photosynthesis is at least 3.3 thousand million years old and that blue-green algae must have populated wide areas on Earth already at that time.'[27]

But this conclusion, in agreement with Barghoorn and Schopf's findings, confronted the scientists dealing with the evolution of the atmosphere with another problem and an apparent contradiction. If photosynthesis is as old as this, oxygen must have been fed into the environment by organisms that gain energy through this process. But the very same Witwatersrand deposits containing the carboniferous substance in which Schidlowski had found his first pieces of evidence for the existence of photosynthetic and hence oxygen-producing organisms more than two thousand million years old, also contain the nonoxidised grains of pyrite and pitchblende which Ramdohr had taken as evidence that there was no oxygen in our atmosphere at that time (see page 32). How can this apparent contradiction be elucidated? Where had the oxygen produced by photosynthesis disappeared to?

Banded iron formations

When blue-green algae and other green plants split water to produce hydrogen for the synthesis of carbohydrates, free oxygen is liberated as a 'waste product'. To the first primitive unicellular organisms, however, this highly reactive chemical element was extremely toxic. Nowadays substances liberating oxygen, such as hydrogen peroxide or potassium permanganate, are used as disinfectants. Thus the transition from hydrogen sulphide consumption to the use of water as a raw material for photosynthesis (see page 44) was possible only provided that the toxic oxygen could be neutralised. This was achieved by a substance dissolved in the oceans at that time.

As long as there was no oxygen in the atmosphere, the Earth's crust mostly contained iron that was not fully oxidised, called bivalent (ferrous) iron in the chemist's jargon. Unlike the fully oxidised trivalent (ferric) iron that exists today, recognisable by its typical rusty colour, compounds of bivalent iron are easily dissolved in water. Through weathering, the continental rocks dissolved and were carried to the oceans. There the iron accumulated, whereas at present only traces (millionth of a per cent) of dissolved iron compounds are found in sea water.

'For oxygen, a solution of ferrous iron compounds is like a sponge,' said Schidlowski. 'Iron immediately seizes the oxygen, even minute amounts of it. Hence it is plausible that oxygen liberated by the blue-green algae via photosynthesis was immediately bound to the bivalent iron dissolved in the oceans. Thereby trivalent ferric oxide was formed, which, being nonsoluble, was deposited in the sea-bed.'

In fact old sediments which once had lain on the bottom of the sea contain huge amounts of iron ore, the so-called banded iron formations. The iron contained in these ores, which supply 70 per cent of the world's iron production, is fully oxidised (trivalent). It was the idea of Preston Cloud, an American geologist, that these deposits were 'waste products' resulting from photosynthesis of primeval blue-green algae. This idea is supported by the fact that most of the banded iron formations originate from a time before the atmosphere contained appreciable amounts of oxygen. According to Cloud, it is very unlikely that the huge amounts of oxygen necessary for the formation of these ores could have been liberated by some nonbiological process.[28]

Schidlowski showed me polished samples of such minerals. Bands of deep dark red, nearly black, that predominantly consist of ferric oxide alternate with light red ones consisting of dispersed iron embedded in a flint-like material. According to Cloud, this banded structure is likely to be due to periodic variations in the intensity of photosynthesis. As yet nobody knows how this might have happened in detail. The thickness of the different layers surely excludes annual variations as a likely mechanism. The banded structure rather suggests variations with considerably longer periods.

Banded iron formations are found in different parts of the world. The volume of these deposits, formed more than two thousand million years ago, is often enormous. 'Until a few years ago sediments found in South Africa dating back 3.3 thousand million years were thought to be the oldest ones,' said Schidlowski. 'It was doubted whether older sediments would be found at all, since sediments are also subject to mountain formation and erosion throughout geological cycles. In 1973, however, British geologists discovered sediments close to the west coast of Greenland with an estimated age of 3.7 thousand million years. A great deal of these also consist of banded iron formations.'[29]

A scholarship provided by the Mainz Institute enabled Peter Appel, a young Danish geologist, to start collecting samples of these rocks. 'The working conditions are extremely difficult there,' Schidlowski continued. 'The deposit lies at the edge of the inland ice. It is only for about two months in summer that this place is accessible at all.'

In September 1976, when I visited the Mainz Institute for the first time, the first shipload of rock samples from Greenland had just arrived in Copenhagen. As a few of these had been sent on ahead by aircraft, Schidlowski could hand one to me.

A stone 3.7 thousand million years of age. It did not look any different from the other banded iron samples that were several hundred million years younger. However, the scientists expected the rock samples from Greenland to reveal important information. 'We are anxious to find out whether these sediments also contain carbon of biological origin,' said Schidlowski.

In the two years that followed, 124 of these rock samples were analysed at Mainz for their isotopic composition of carbon. The enrichment of carbon-13 measured amounted to about half that usually found in young sediments. 'This alone would be a strong indication for the existence of living organisms which, 3.7 thousand million years ago, were already capable of operating photosynthetically', stated Schidlowski, 'because it is unlikely that such large amounts of reduced carbon could have been produced through nonbiological processes. Yet the evidence is in fact even stronger. Geological analysis has shown that after these sediments were formed, they must have been subjected to different geological processes that have altered the isotopic composition of the carbon. If this is taken into account, one comes up with the surprising result that these old sediments from Greenland originally contained the same fraction of biologically produced carbon as is found in young sediments, about one fifth of the total carbon. We therefore have to get used to the idea that life on Earth is far older than previously assumed, and that eight hundred million years after the formation of our planet, there was already a flourishing biosphere. It seems that some sort of ancestor of blue-green algae existed at that time on photosynthesis.'[30]

'Besides your isotopic analysis, are there any other pieces of evidence for the existence of life as early as 3.7 thousand million years ago?' I asked.

At the time I asked this question pertinent research had only just started, but now results are coming in. Using the very sensitive technique of vaporisation gas chromatography which can find evidence for admixtures in the parts-per-billion range Jürgen Hahn did spot decomposition products of chlorophyll. Moreover the palaeontologist professor Hans D. Pflung of the University of Gießen, when examining samples of the Greenland sediments with the electron microscope, found structures which he thinks are unicellular fossils.[31]

According to Preston Cloud's theory, the ample abundance of banded iron formations in the Greenland sediments can be taken as further evidence for blue-green algae having been involved in the formation of these deposits. If this theory is right – and the scientists at Mainz tend to regard this as likely – one can easily understand why the Witwatersrand deposits contain, within the same sediment layer, evidence for an atmosphere without oxygen and for the existence of organisms living on photosynthesis. For about two thousand million years the oxygen produced through photosynthesis by the blue-green algae was immediately bound by dissolved bivalent ferrous iron, and did not escape into the atmosphere.

In the course of this very long period of time, certain species of algae evolved defensive enzymes to counter 'poisonous' oxygen. But only when the stock of bivalent iron dissolved in the oceans was nearly exhausted, and the fresh supply from the continents was insufficient to cover the requirements, did these defensive mechanisms provide a better chance of survival. Those organisms possessing the right enzymes could thrive in waters deficient in ferrous iron, where oxygen accumulated.

Now oxygen dissolved in the water began to escape into the atmosphere. It could thus also attack the bivalent iron contained in continental ores and convert them, in the course of the weathering process, into insoluble trivalent ferric oxide. The red sandstones thus formed on the continents date back up to some 1.8 thousand million years. It is most likely that from this time on, appreciable amounts of oxygen accumulated in the near-surface layers of the atmosphere.

With the oxidation of continental rocks, the fresh supply of bivalent iron into the oceans was gradually cut off. Sediments that are one to two thousand million years old seldom contain banded iron formations. Their volume does not reach that of the older deposits. In sediments younger than one thousand million years no banded iron formations are found at all. The types of algae that could only thrive in oceans rich of bivalent iron and devoid of oxygen had thus been driven away by other oxygen-resistant species.

From then on, the oxygen liberated through the photosynthesis of these plants was no longer bound by substances dissolved in the water. For the most part it escaped into the atmosphere. But since the oxidation of bivalent iron is much slower for the continental rock weathering process as compared to that of iron dissolved in water, oxygen accumulated in the atmosphere, until finally a new equilibrium was reached. During the course of the Earth's history, a total of about ten times the amount of oxygen presently found in the atmosphere has been consumed in the oxidation of bivalent iron.

As a result of photosynthesis, new 'artificial' environmental conditions were created on Earth that were maintained by the activity of living beings. The accumulation of oxygen in the atmosphere fundamentally changed the geological conditions. Everything oxidisable thus became gradually oxidised when exposed to the air: carbon, sulphur, iron and many other substances.

The radiation climate was also changed completely. As long as the atmosphere contained no oxygen and thus the ultraviolet light could penetrate to the ground without being attenuated, the expansion of life was limited. According to Berkner and Marshall, life could only exist at that time under at least 10 metres of water, since most of the lethal ultraviolet radiation was absorbed in such a depth. The calcite or silica plating of many species of blue-green algae and other biogenetically very old beings is likely to have originally served as a radiation shield. In fact, blue-green algae are still extremely resistent to ultraviolet radiation which was of course much stronger when these organisms evolved.[32]

With the gradual accummulation of oxygen in the atmosphere, ozone was formed that absorbed ultraviolet radiation. According to Berkner and Marshall, when the atmospheric oxygen level had reached only 1 per cent of its present concentration, the ozone shield was already dense enough to enable life under only a few centimetres of water. It is generally accepted that this oxygen level was reached about 1.5 thousand million years ago.

The living organisms experienced completely new biochemical conditions created by the gradual buildup of oxygen in the atmosphere. Environments protected from oxygen, still very poisonous to many species, became sparser. This trend continued until the situation that prevails today was reached, where few anaerobic environments exist. Such an environment is where oxygen-consuming substances prevail, such as decaying organic matter. Only in such an environment, in the depth of the soil, in swamps, in certain regions of the sea-bed, or in the digestive tract of animals, can anaerobic microorganisms still survive.

For the new species, resistant to oxygen, the changed environmental conditions implied a new chance. As soon as the oxygen reached about 1 per cent of its present atmospheric level, a mechanism considerably more efficient than fermentation became possible for liberating the chemical energy stored in carbohydrates – that of oxygen respiration. It is reasonable to assume that this transition occurred at this very oxygen level. There are bacteria, so-called facultative anaerobes, capable of producing energy without oxygen by fermentation as well as by respiration in the presence of oxygen. Louis Pasteur, the famous French scientist, had already found that the transition from one form of energy production to the other occurs at about 1 per cent of the present oxygen level in the atmosphere.

For a given amount of carbohydrate, an organism can liberate about nineteen times as much energy by respiration than by fermentation.[33] Thus respiring organisms are considerably more efficient than fermenting ones. This is likely to be the answer to the question as to why, after the relatively fast evolution of life and photosynthesis, about two thousand million years of a 'blue-green algae age' had to pass without any spectacular biological development. The oceans had first to be deprived of their bivalent iron, which was the necessary condition for the buildup of oxygen in the atmosphere and hence the origin of respiration, before the evolution of higher organisms could start.

All living beings differing from primitive bacteria and blue-green algae by having a cell nucleus meet their energy requirements through respiration. These higher organisms, the so-called eukaryotes, are likely to date back about 1.4 thousand million years, as traces found by J. W. Schopf in Australian sediments indicate. In a joint paper with Dorothy Oehler, he points out that, on the average, cells of eukaryotes are considerably larger in diameter than those of bacteria or blue-green algae, and that traces of such larger cells are found in most sediments younger than 1.4 thousand million years.[34]

It appears as if this newly developed structure of the unicellular beings was the precondition for the next major stage of evolution, which was the origin of multicellular organisms. All multicellular beings are eukaryotes and thus respiring creatures. Their traces date back nearly seven hundred million years. From that time a vigorous evolution of continually newer and higher forms of life began.

The continents were still not inhabitated then. For the living beings that had all developed in the oceans, it was difficult to advance to a non-aquatic environment, where special devices to provide water were required. Moreover, gravitation posed a new problem to the multicellular beings and the radiation climate on land was still extremely harsh. With the gradual increase in atmospheric oxygen however, more and more ultraviolet radiation was absorbed at higher altitudes, and the ozone layer gradually shifted upwards. When about 10 per cent of the present atmospheric oxygen level had been reached, according to Berkner and Marshall, the conditions were right to populate the continents.

It is probably not very realistic in this context to talk of a 'conquest of the land'. We are not dealing with some sort of ancestors of the Vikings who, bold and full of venturesomeness, set out for new frontiers – and we leave the question open as to whether this was the case with the real Vikings – but rather with ancestors of some sort of Robinson Crusoe, i.e. with organisms forced, due to unfavourable conditions, to adjust to a new environment. Inhabitants of gradually drying lakes or lagoons, and organisms thrown ashore by the tides, had before been doomed to destruction by ultraviolet radiation. Now there was a slight chance of survival, or more precisely, among thousands of millions of creatures facing such unfavourable conditions, now and then one would somehow survive, thanks to a favourable coincidence of just the right mutations, and thus become the ancestor of a certain species capable of living on land. That way the continents gradually became colonised by microorganisms and plants, during the geological epoch called silurian, some 420 million years ago. At the beginning of the carbon age, about 350 million years ago, huge multicellular plants existed on land. This new land vegetation with its photosynthesis further increased the production of oxygen.

Before plants grew on the land, the bare continental rocks were unprotected and therefore exposed to the impact of solar radiation, winds and weather. Weathering process and mountain erosion, and hence geological loss of oxygen, must have been much more rapid then than under present conditions. Thus the newly formed plant cover contributed to an increase in the atmospheric oxygen concentration not only directly through the enhanced rate of photosynthesis, but also indirectly through a reduction in mountain erosion. Moreover, the organisms decomposing the remnants of dead plants were still evolving. Consequently, the decay processes of organic matter were certainly slower than at present. The huge coal deposits of the carbon age are probably just a few remnants of enormous amounts of dead plant material that accumulated at that time. This explains why Berkner and Marshall, and other scientists, assume that the atmospheric oxygen level around that time increased by a factor of ten within only seventy million years, to reach or perhaps even exceed the present level.

It is disputed whether, and to what extent, changes in atmospheric composition have occurred since then. From the uninterrupted evolution pattern of land organisms one can certainly conclude, however, that certain minimum conditions for their existence have always prevailed. Hence the carbon dioxide content in the atmosphere has probably never been lower than one third of its present level, because otherwise photosynthesis would have come to a standstill. On the other hand, the evolution of the respiratory organs in animals clearly reflects the gradual increase in the oxygen content of the atmosphere. Step by step, more efficient forms of energy use have developed that required an increasing consumption of oxygen. Warm-blooded animals, that need at least two thirds of the present atmospheric oxygen level, have been living on Earth for about 250 million years. Hence one can assume that the oxygen concentration has not fallen below this limit during this time.

The origin of biological cycles

Since the 'invention' of photosynthesis, there have been two classes of living beings on Earth; namely self-supporting and non-self-supporting organisms. From their environment, the self-supporting (autotrophic) creatures take up merely building materials such as compounds of nitrogen, sulphur, and phosphorus and the basic substances for photosynthesis, such as carbon dioxide and water (or hydrogen sulphide). The biogenetically older non-self-supporting (heterotrophic) beings do not only need building materials for their metabolism, but also fuel, basically carbohydrates, which they cannot produce themselves. In the beginning they lived on the various compounds rich in energy, which were dissolved in the primordial soup and to some extent replenished as long as the atmosphere did not contain oxygen. With the origin of photosynthesis, a new source of nutrients was made accessible to them: the metabolic and decomposition products of the autotrophs, and finally the autotrophs themselves, all containing carbon compounds with stored chemical energy.

Thus, in principle, the conditions for starting a biological cycle were attained. Carbohydrates built up by the autotrophs could be decomposed again by the heterotrophs. Before respiring organisms had evolved however, this cycle was not closed, since fermentation is not an oxidation process capable of reproducing the basic substances carbon dioxide and water from carbohydrates. Hence only a small part of the energy required for carbohydrate formation is liberated in the process. The end-products of fermentation are still relatively complicated oxidisable carbon compounds.

Under favourable conditions, deposits of fossil fuel can be formed from the partly decomposed remnants of dead organisms. In general however, the volatile compounds escape, and only pure carbon remains dispersed in the sediments. During the two thousand million years when photosynthesis existed but free oxygen and respiration did not, huge amounts of carbon accumulated in the sea-bed and thus entered the geological cycle in that way.

It appears as if the first fairly closed biological cycles were achieved by bacteria that gain their energy through neither fermentation nor respiration in the usual sense. Rather they maintain a metabolism whereby oxygen for the oxidation of decaying organic matter is taken up from certain compounds rich in oxygen, such as nitrates. In part 4 of this book we shall come back to the so-called nitrate respiration. Besides nitrates, sulphates and trivalent ferric oxide can also serve as an oxygen source. The end-products of this metabolism are carbon dioxide and water, and reduced substances such as nitrogen, sulphides or compounds of bivalent iron, respectively. Such bacteria are still found today in the bottom of the ocean and in the depths of the soil. They are anaerobic in a facultative or obligatory sense, i.e. they do not necessarily require free oxygen (facultative) or they cannot stand it (obligatory). They still play an important role, often ignored, in the global circulation of substances. Yet their importance was certainly much greater during the 'blue-green algae age', before the origin of an oxidising atmosphere, when they presumably populated the oceans too. Had such microorganisms, reconverting trivalent iron into soluble bivalent iron, not existed, it probably would have required a much shorter time than two thousand million years for all the bivalent iron dissolved in the oceans to be used up, and hence oxygen could have escaped into the atmosphere much earlier.

Sulphur and carbon are likely to have been linked in another cycle. On one side of the cycle purple bacteria (see page 43) produced carbohydrates and sulphates from hydrogen sulphide and carbon dioxide, while on the other side, bacteria that can be classified as 'sulphate respirators', burnt the decaying organic substance (carbon) to carbon dioxide using the oxygen from the sulphates.

With the origin of an oxidising atmosphere, the environments where such reducing microorganisms could thrive became very limited. Thus photosynthesis and respiration, supplementing each other, became by far the most important biological cycle. Both are complicated processes passing through various intermediate stages. If, however, only the basic substances and end-products are considered, they can be simply described by the following equilibrium equations:

$$\text{carbon dioxide} + \text{water} + \text{energy} \underset{\text{respiration}}{\overset{\text{photosynthesis}}{\rightleftharpoons}} \text{carbohydrates} + \text{oxygen}$$

$$CO_2 + H_2O + \text{energy} \rightleftharpoons (CH_2O) + O_2$$

Herein CH_2O put in parantheses does not stand for formaldehyde, but rather for the ratio of the carbon (C), hydrogen (H), and oxygen (O) atoms that form the basic building block of larger molecules, such as glucose ($C_6H_{12}O_6$).

This reaction scheme of photosynthesis does not show that the total amount of oxygen liberated in this way does not originate from the carbon dioxide but rather from the water. For the subject of our book though, this is not important. We can confine ourselves to the statement that every molecule of carbon dioxide taken up through photosynthesis by green plants, corresponds to one oxygen molecule released into the environment, and one atom of carbon entering the biological processes.

Is the long-term cycle balanced?

Since the evolution of respiring organisms, most of the oxygen liberated through photosynthesis is consumed by them. Those unoxidised amounts of carbon that were buried in the sediments and the corresponding 'surpluses' of oxygen in the atmosphere (see pages 3 and 20) have probably not always been as small as at present. Compared to the total production however, they have always only constituted a negligible fraction throughout the last three hundred million years.

Some of the oxygen surplus oxidises the carbon contained in the sediments which, in the course of geological cycles, after hundreds of millions of years, gets back to the earth's surface. Another fraction is used up in weathering processes oxidising iron, sulphur, and other substances. Reverse reactions, whereby fully oxidised compounds of iron and sulphur are reduced, take place mainly in the ocean-bed and in the Earth's interior.

Thus there are several groups of processes, not directly dependent on each other, that can be put together as a long-term cycle. During the past few years scientists have tried repeatedly to quantify this cycle and to construct a flow diagram that comprises at least its most important features.[35] Figure 7 is an example published by the Mainz group. As far as the details of the various branches and the figures are concerned, the models published so far differ quite considerably. Despite the different figures however, all authors base their models on the assumption that there is some sort of stationary equilibrium, or at least was, before man interfered with the natural cycles.

The maintenance of such a stationary state is possible only if the oxygen surplus, due to the sedimentation of unoxidised carbon, exactly balances the oxygen loss through geological processes. It appears as if this assumption is not in conflict with observed facts.

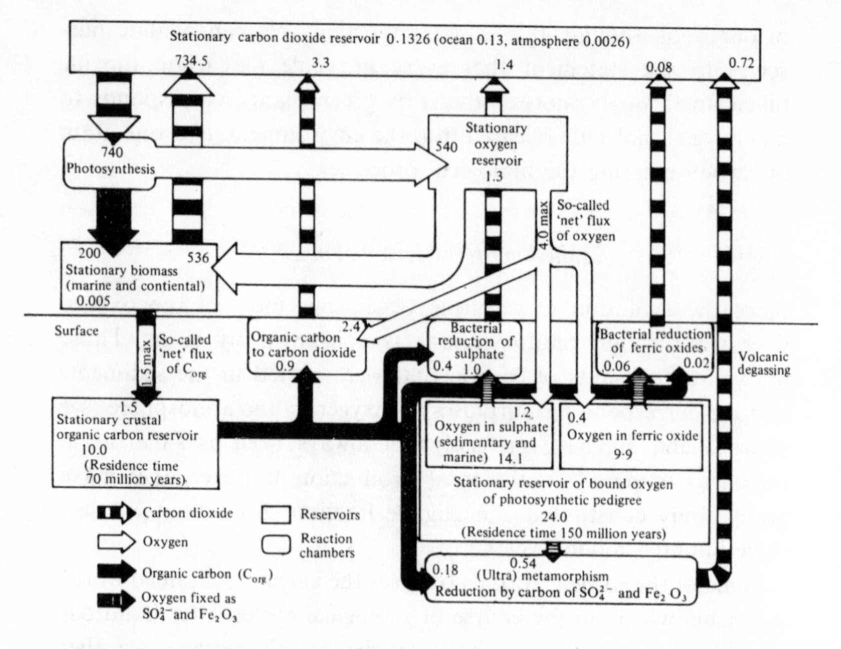

Figure 7 Chart showing fluxes in the combined organic carbon–oxygen cycle according to the model formulated by the Mainz group. Figures for reservoirs (framed) are in thousand billion tonnes, those for fluxes in hundred million tonnes per year. The carbonate (limestone) cycle comprising about four times the carbon involved in the 'organic' cycle is not shown here. According to Whittaker and Likens, the budget for the photosynthetic cycle (wide channels) is nearly four times as great as that given here, i.e. nearly eighty thousand million tonnes of carbon per year. This is however of no direct consequence with respect to the volume involved in sedimentary cycle (narrow channels). The figures for the so-called 'net' oxygen flux and 'net' flux of organic carbon (C_{org}) are maximum values. According to the Mainz group the actual figures might be smaller by a factor of 1.5 to 4 – which would correspond to the figures assumed by Garrels and Mackenzie. The residence times in the sedimentary reservoirs would then be correspondingly longer. The stationary equilibrium of the cycle assumed here has not yet been proved. (From Schidlowski *et al.*, *Precambrian Res.*, **2**, 55, 1975.)

Independent estimates, published in the scientific literature, yield matching magnitudes for both processes. James C. G. Walker however, an American geologist and author of one of these flow diagrams, expressly points out that the data available at present are far too incomplete to be able to conclude with confidence that such an equilibrium is indeed being maintained.[36]

There is no obvious direct relationship that matches the amount of carbon getting burried in the sediments (in the fourth reservoir) with the geological loss of oxygen. As a matter of fact such an equilibrium *can not* have existed in the geological past. If there had not been a period when the liberation of surplus oxygen exceeded the rate of oxygen loss through weathering processes, no oxygen reservoir could have been built up in the atmosphere.

A growth in the abundance of atmospheric oxygen can only be expected if a corresponding increase in the amount of carbon stored in organic matter or in the sediments occurs, and such an increase can persist only as long as sufficient raw material, namely carbon dioxide, is available, from which green plants can produce oxygen. Originally this carbon dioxide was exuded from the Earth's interior (see page 29), but these outgassing processes were to all intensive purposes terminated some three hundred million years ago.[37] Most of the gases released in volcanic eruptions today are probably recycled materials: reaction products of sediments which were transported into the Earth's interior through large-scale geological processes.

When the outgassing from the Earth's interior was essentially at an end, the most important condition for an equilibrium state probably prevailed: from this time on, the total mass of circulating substances did not further increase. The only changes possible were such that the content of one reservoir could be increased at the expense of another. When essentially no *new* carbon dioxide emanated from the Earth's interior, the amount of atmospheric oxygen could increase only at the expense of the amount of oxygen stored in one of the other reservoirs, for instance in carbon dioxide or other substances containing oxygen.

For short periods of time, an increase in atmospheric oxygen through an intensified rate of photosynthesis is possible when the amount of carbon entering the sediments is correspondingly increased. However, such a growth process can only be maintained as long as the higher rate of consumption of atmospheric carbon dioxide is compensated for by the subsequent delivery from other reservoirs. Model calculations carried out by the American geologists Garrels, Mackenzie and Lerman have shown, for instance, that doubling the rate of photosynthesis would bring this process to a standstill within a period of only one million years due to a shortage of atmospheric carbon dioxide.[38] Within this time-span, considered short in geological terms, the concentration of atmospheric oxygen would increase from 21 to 22 per cent. A continuous growth over longer periods is possible only at much slower rates that depend on how fast the used atmospheric carbon dioxide is replenished from the oceans (and ultimately from the carbonates). The gas exchange processes between the deep ocean layers and the atmosphere simply could not keep up with the tremendous carbon dioxide consumption that would be the likely consequence of a doubling of the global rate of photosynthesis. Moreover, other feedback mechanisms counteracting an oxygen increase are likely to exist, although their exact form is still under dispute.

Another condition for maintaining an equilibrium state demands that the total amount of oxygen bound in iron and sulphur compounds does not increase, in spite of the fact that large masses of oxidisable iron and sulphur compounds are still available in the Earth's crust capable of binding far more oxygen than is at present in the atmosphere. Thus the oxidation reactions that certainly take place in the course of rock weathering have to be compensated for by reverse processes. These are maintained through the activity of reducing microorganisms, such as sulphate respirators (see page 60), as well as by chemical reactions in the Earth's interior, when sediments are molten at greater depths (geologists call this latter process 'granitisation' or 'ultrametamorphosis'). In both cases the reduction occurs in the presence of sedimentary carbon which thereby is converted to carbon dioxide. The amount of oxygen consumed by weathering processes, and the production of carbon dioxide through reducing biological and geological processes need not necessarily balance each other exactly. Possibly feedback mechanisms exist, but they are not known at present.

The atmospheric oxygen concentration cannot increase directly on account of the oxygen chemically bound to iron and sulphur since, due to the low temperatures on the Earth's surface, there are no reactions that could liberate free oxygen from such compounds. The product of the reduction processes is carbon dioxide. Only as a second step can oxygen be produced through photosynthesis by plants. On the other hand, atmospheric oxygen can directly be used up in the formation of compounds of iron and sulphur. If, for instance, due to an increased rate of erosion, the geological loss of oxygen becomes accordingly larger, this may lead to a reduction in atmospheric oxygen, unless the biological or geological reduction of sulphur and iron compounds, rich in oxygen, increases at an equivalent rate, Moreover, increased erosion does not only lead to an increased oxygen loss due to reactions with iron and sulphur, but also to an increased oxidation of carbon contained in the sediments, and hence a reduction in the amount of carbon stored in the fourth reservoir. Thus atmospheric oxygen is as a consequence further reduced.

According to Garrels, Mackenzie and Lerman this is exactly the situation we are faced with at present. 'There is one important point wherein our flow diagram and that presented by Schidlowski and his colleagues differ; while the Mainz group postulate that a stationary equilibrium exists under *present* conditions, we are doubting that,' Fred Mackenzie told me during the Dahlem conference. 'In their model, the geological loss of oxygen is fully compensated for by the oxygen surpluses due to the sedimentation of carbon, when present erosion rates are taken into account. We believe, that such an equilibrium can only be assumed as an *average* condition over the last three hundred million years. At the present time however, due to man's interference in these natural cycles and thus increased erosion rates, this equilibrium is likely to be disturbed.'

Schidlowski readily admitted that this may be the case. 'Our flow diagram was just a first attempt to see if we could somehow come to grips with these complicated processes. One cannot exclude the possibility that short-term variations, such as those due to human interference, are superimposed upon the longer-term equilibrium. The arguments presented by Garrels and his colleagues are plausible indeed.'

The global extent of erosion, and hence the geological loss rate of oxygen as well, depends on the dimensions and average heights of the continents. These have been subject to various changes throughout the Earth's history. The erosion rate of young high-folded mountain ranges is much faster than that of old massifs that have been largely levelled already. Thus, during periods following enhanced mountain formation, the erosion rate must have been considerably greater as well. In the long-term average however, the system has always been balanced and has not deviated too far from the equilibrium, says Mackenzie.

The growth of the vegetation cover on the continents during the last three to four hundred million years is an additional factor that has certainly slowed down the rate of erosion and has therefore led to a new equilibrium state. This condition is now being disturbed by man-made changes in the vegetation cover. Based on the figures of Judson (see page 16) that civilisation has trebled the erosion rate, Garrels, Mackenzie and Lerman have assessed the long-term consequences of such conditions with respect to the atmosphere. Their results indicate that within two million years the atmospheric oxygen level will have decreased by some 15 per cent, while the carbon dioxide content will have doubled.[39]

'For the sort of time-spans we are concerned with here at the Dahlem conference, this is of no importance at all,' Mackenzie remarked. 'Even if Judson's figures are still too small, as we think is possible, that is, if the present erosion rate were four or even five times the average that it has been over the past several hundred million years, this would have no dramatic effect on the composition of the atmosphere for the next hundreds or even thousands of years. In the long-term, however, this man-made increase in the erosion rate, and hence the geological loss of oxygen, will have an even more drastic impact than the burning of fossil fuels.'

3

Carbon dioxide increase – how great is the risk?

Atmospheric carbon dioxide and fuel consumption

From 1957 to 1975 the amount of carbon dioxide in the air has increased from 312 to 326 parts per million (ppm), almost 5 per cent. This trend has been measured both by American scientists under C. D. Keeling since 1957, and by Swedish meteorologists led by B. Bolin since 1963 (see page 5) in areas far away from any localised source of carbon dioxide so that the background and thus the worldwide trends in carbon dioxide concentration might be observed.[1] The Americans collected data on Mauna Loa, Hawaii, and at the South Pole while the Swedes made their measurements on the air-intake systems of commercial aircraft flying the polar route.

For the period before 1957 there are only odd unsystematic and hence somewhat unreliable data from various scientists which have been collected by the English geophysicist G. S. Callendar.[2] Though showing considerable scatter, they are in general consistent with the later, more precise observations.

The concentration of carbon dioxide in the air around 1860, before the effects of industrialisation were felt, is assumed to have been about 290 ppm. In the hundred years and more since then, the concentration has increased by about 30 to 35 ppm, that is by 10 per cent. This is equivalent to 235 to 275 thousand million tonnes of carbon dioxide. The present rate of increase is approximately 8 thousand million tonnes, or 1 ppm, per year.

To meteorologists and geophysists making these measurements it seemed at first 'obvious' that this extra carbon dioxide is the result of the combustion of fossil fuel.[3] Consequently, data on the amounts of fossil fuels consumed were collected by C. D. Keeling so that they might be compared with the carbon dioxide data.[4] The tedious part of this research was to assemble the statistics from a

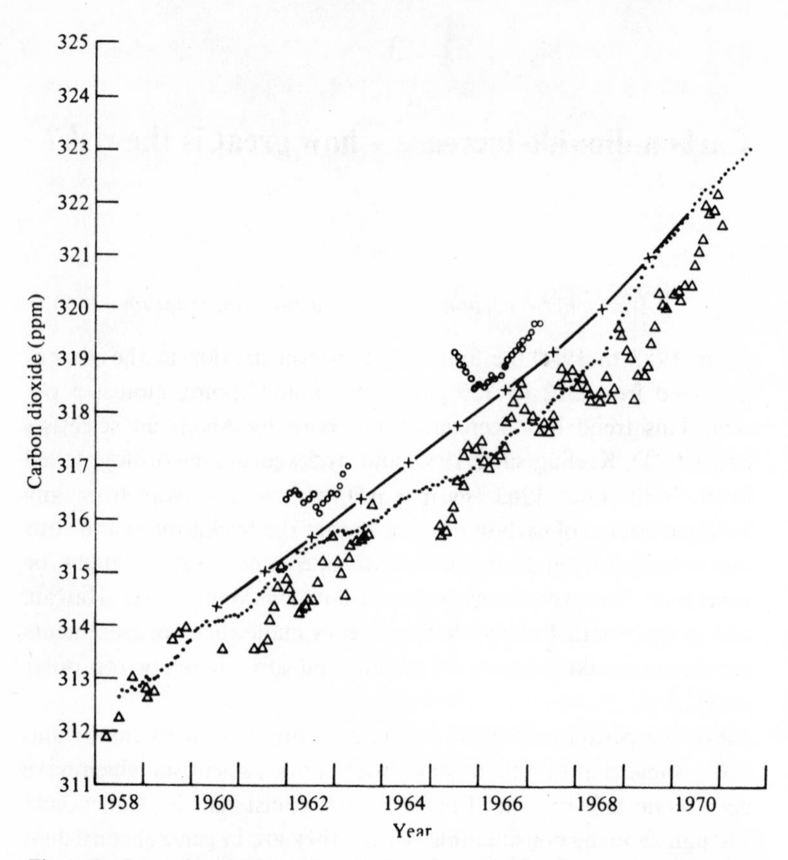

Figure 8 Increasing trend in atmospheric carbon dioxide concentration, according to different sources (see also figure 2, page 4). + = Scandinavian aircraft flasks, adjusted to 1 July; ● = Mauna Loa, Hawaii 12-month running means; △ = South Pole; ○ = Point Barrow, Alaska, twelve-month running means. (From L. Machta in Hess (ed.), *Weather and Climate Modification*, p. 693, Wiley, New York, 1974.)

whole variety of sources dating right back to 1860, to check and compare each piece of data to make sure it was accurate and realistic, and then finally draw up a table of this data. The statistics that exist for the years before 1860 are rather inaccurate and can be neglected anyway since the amount of fuel concerned was very small.

The statistics themselves do not list the amount of carbon dioxide produced, but rather the energy consumption in coal equivalent units. Since some of the fuels used were hydrocarbons (oil, natural gas etc.) which, when burnt, produce energy by oxidation of their hydrogen as well as of their carbon content, the amount of carbon dioxide produced in this process can only be found in complicated calculations. Keeling also noted two further sources of carbon dioxide, each accounting for about 2 per cent of the release into the atmosphere: flaring (burning) of natural gas in oil fields and cement production.

The burning of wood, animal dung, straw etc., which constitutes about half the total energy source in developing countries,[5] was not included by Keeling and nor were the considerable amounts of carbon dioxide released by slash-and-burn agriculture, the burning of stubble after harvest and natural forest fires.[6]

In terms of methodology it is justifiable to distinguish between these latter sources and the combustion of coal, oil or natural gas. Fossil fuel consumption is simply an accelerated form of the long-term geological cycle. It removes carbon from the fourth reservoir (see page 5) with a consequent increase of atmospheric carbon dioxide and a (very small) decrease in the atmospheric oxygen concentration. The burning of wood and plant matter, on the other hand, is a substitute for the short-term biological cycle. As long as no more wood is burnt than is reproduced by the forests, no change in atmospheric carbon dioxide concentration will result. Burning in this case just replaces the natural rotting of wood. It is only when forests are cleared at a greater speed than they can regenerate, and thus the world's biomass as a whole shrinks, that the concentration of carbon dioxide in the air increases (see page 7).

It is not clear whether Keeling fully appreciated this when he published his paper or whether he simply neglected the burning of wood etc. from his calculations because there are no reliable data on it and he underestimated their importance. Consequently, from this point of view, his tables are incomplete. Nevertheless, they provide a very important base for an insight into this problem. One sees that between 1860 and 1910 the fossil fuel consumption increased at an almost constant rate of 4.3 per cent per year. Such an increase is called exponential growth since when there is the same percentage increase each year it is like compound interest, with the actual mass of material being consumed each year increasing. Between 1910 and 1945, a time-span containing two world wars and the economic depression of the 1930s, the growth rate is very irregular, amounting to an average value of 1.5 per cent per year. However, between 1945 and 1970, the growth rate was again exponential, this time with an average value of 4.8 per cent per year. Extending this curve and including the oil crisis of 1973, it is clearly evident that the trend has now slowed down again.

According to Keeling's calculation for the period 1860 to 1970, a total of between 360 and 460 thousand million tonnes of carbon dioxide entered the atmosphere as the result of burning fossil fuels. Though this is not a very precise figure there is no doubt that it was considerably more than the amount which actually remained in the atmosphere (235 to 275 thousand million tonnes of carbon dioxide). The picture is even more clearer if one considers just the recent past. In 1975, 18 thousand million tonnes of carbon dioxide (equivalent to 5 thousand million tonnes of carbon) were released into the atmosphere but the atmosphere showed an increase of only

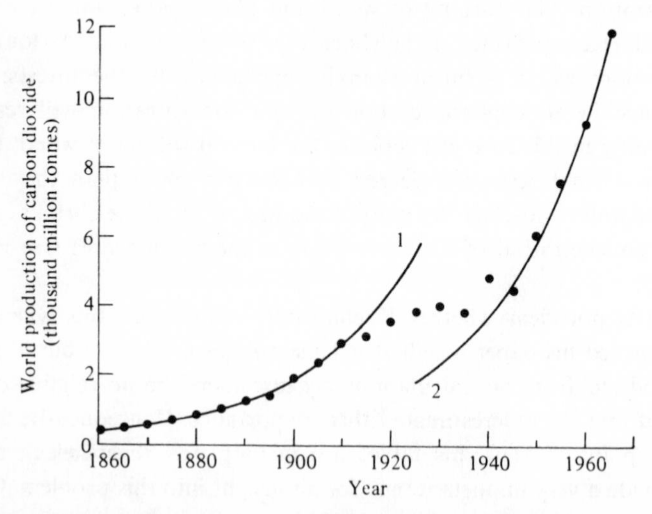

Figure 9 Worldwide production of carbon dioxide from burning of fossil fuels. Curves 1 and 2 are exponential fits with annual growth rates of 4.3 per cent and 4.8 per cent, respectively. (After data of Keeling; from D. R. Kester and R. M. Pytkowicz in W. Stumm (ed.), *Global Chemical Cycles and their Alterations by Man*, Berlin 1977, p. 101.)

some 8 thousand million tonnes (equivalent to 2.2 thousand million tonnes of carbon). The question therefore arises: Where is the rest of this carbon dioxide going to?

The most straightforward answer to this question is: into the oceans. The ocean waters contain about sixty times as much carbon dioxide, mostly as bicarbonate ions, as the atmosphere. There is a continual gas exchange between the atmosphere and the oceans and, together with chemical reactions that take place between the oceans and the minerals on its floor, an equilibrium is established. If the equilibrium is disturbed by externally increasing the concentration of carbon dioxide in the air, then the oceans would absorb more and more carbon dioxide (see page 30). In the long run, practically all the carbon dioxide released by the burning of fossil fuel would end up in this ocean reservoir. What we see however is that in the last hundred years, the increase in carbon dioxide release has been so great that the oceans have been unable to keep pace. The reason for this is because it takes a relatively long time to achieve a complete mixing of the oceans (approximately 500 to 1500 years). Generally the oceans have a warm top layer, some hundred metres thick, which is in close contact with the atmosphere. The rest of the ocean waters, on average several thousand metres thick, and therefore constituting the bulk of the ocean, have no direct gas exchange with the atmosphere and hence atmospheric carbon dioxide. Only in polar regions, where the surface waters too are cold and thus the deep waters can circulate to the surface, is a direct gas exchange possible.

The carbon dioxide equilibrium is not only dependent on the relative air–water concentrations, but also on the chemical composition of the seawater and the temperature. The colder the water the more carbon dioxide can be absorbed by the oceans. Under normal undisturbed conditions, warm seawater in the tropics releases carbon dioxide into the air while in polar regions carbon dioxide is absorbed by the oceans from the air. The carbon dioxide produced through fossil fuel consumption must enter this transport cycle.

There is another cycle too which is of significance in this context. Carbon dioxide dissolved in rain water or ground water forms carbonic acid and can dissolve limestone. Thus water-soluble calcium bicarbonate is formed which is transported by the rivers to the oceans. The world's oceans are therefore saturated with calcium bicarbonate which is absorbed by many forms of life to form their shells or protective coating and is then deposited as limestone sediments. In the processes of shell formation carbon dioxide is released again first into water and then into the atmosphere, depending on the state of the atmospheric–oceanic equilibrium.

The fact that carbon dioxide moreover is absorbed in great quantities in photosynthesis and released again in respiration or decay was discussed in detail in part 1. To trace the path of fossil fuel carbon dioxide in this confused set of interacting cycles is not at all easy. We are assisted however by the fact that the carbon dioxide produced by burning of fossil fuels bears a special label distinguishing it from atmospheric carbon dioxide.

Eighty tonnes of labels

In the upper layers of the atmosphere, cosmic rays constantly transform small amounts of nitrogen into the isotope of carbon with atomic weight 14 – not to be confused with carbon-13 which was discussed in part 2 (see page 46). The annual production amounts on average to about 9.8 kilograms and altogether there are about eighty tonnes of carbon-14 on the Earth. This is not a printing error. We are not dealing with millions or thousands of millions of tonnes as we have been when considering most other quantities in this book. There really are only eighty tonnes. Carbon-14 is a radioactive isotope which in a relatively short space of time, geologically speaking, is transformed back into nitrogen through the decay of its atomic nucleus, and consequently there is no large accumulation.

The radioactive decay of the atomic nucleus follows a pattern that can be described by a statistical law. One can predict how long it will take for half the amount of a radioactive substance to be transformed into another element. At the same time its radioactivity halves. In the case of carbon-14, this so-called half-life is 5568 years. Radioactive carbon is a so-called beta-ray emitter since its atomic nucleus expels an electron as it decays. This electron carries a fixed amount of energy with it and thus can be distinguished from electrons emanating from other beta-ray emitters.

The 'radio carbon' in the upper atmosphere reacts with atmospheric oxygen to form radioactive carbon dioxide and through atmospheric mixing processes is brought close to the ground. The plants absorb it during photosynthesis, and carbon-14 becomes part of the organic material produced. This organic material is consumed by humans and animals alike as food and consequently becomes part of their organic substance. The amounts of carbon-14 in a living organism are minimal, about 0.3 billionths (10^{-12}) of its total amount of carbon. However this radioactivity can be measured precisely, even in such small amounts. Thus atmospheric carbon dioxide (because of its carbon-14 content) bears a radioactive label which is transmitted to all living organisms and other reservoirs that are involved in a continual gas exchange process with the atmosphere.

Where radioactive carbon dioxide is not constantly renewed, for example in the depths of the ocean, the intensity of its radiation is gradually reduced according to its half-life law. By measuring the radiation from carbon dioxide dissolved in waters taken from deep in the ocean, it is possible to establish how long the gas has been in the water and how quickly it has circulated to these depths.

By the same method, it is possible to estimate the age of a bone or a piece of wood found in some archeological excavation. As soon as metabolism ceases and hence no more carbon-14 is absorbed, the clock of radioactive decay starts to tick. With wood, this occurs soon after its formation, since it is then no longer involved in metabolic processes. The oldest wood inside an old tree therefore contains measurably less carbon-14 than the youngest tree-rings just under the bark.

Fossil fuels practically no longer contain any carbon-14 at all. After 55 000 years (ten half-lives) the original amount (0.3 billionths of the total carbon content) has decreased to about one thousandth of this value ($2^{10} = 1024$), after 110 000 years to one millionth of this value, and after 165 000 years to one thousand-millionth and so on. It is known that fossil fuels have been lying in the Earth for some tens to hundreds of million years. The carbon dioxide formed by the burning of these fuels consequently contains no carbon-14. Its label is therefore its lack of radioactive molecules.

The release into the atmosphere of ever-increasing amounts of industrial carbon dioxide has resulted in a measurable dilution of the radioactive carbon content of the atmosphere. This is called the Suess effect, named after its discoverer.[7] It has been possible to establish what was the normal concentration of radioactive carbon before the industrial revolution by measuring its content in tree-rings. It was found that atmospheric carbon dioxide in the 1950s had about 2.5 per cent less carbon-14 than a hundred years before.

Since 1954, however, the carbon clock has been disturbed by the introduction of neutron rays into the atmosphere from hydrogen bomb tests. These bombs produce huge quantities of radioactive carbon and this consequently changed the relationship that had existed up to that time. The Suess effect has thus been overshadowed by the bomb-produced radioactive carbon, whose changes can now be studied in their own right and provide sources of additional information.

By comparing these various observations, both one with the other and with other known natural phenomena, it is possible to try to formulate a total picture of the circulation patterns of carbon dioxide and develop a mathematical model. Such a model uses data on the size of the various reservoirs, the residence times of carbon in the reservoirs, the lifetime of the radioactive carbon contained in these gases, and the exchange ratios between these reservoirs. Of course models are only simplified representations of a much more complicated reality. Nevertheless they can be useful in tracing the ramifications of a complicated set of natural cycles and their variations with a reasonable degree of accuracy, both to account for past events and to predict future trends.

Generally one can build a useful and realistic model by feeding equations and data into the computer and making alterations to certain parameters until it can correctly 'forecast' known past events. In our case, the model must use the atmospheric carbon dioxide content, take into account the Suess effect and other important associated facts. Several scientists have tried such computations with different mathematical models and they have all come to the conclusion that realistic results can only be obtained if one assumes that fossil fuel carbon dioxide is absorbed not only by the oceans but also by land vegetation. The following table shows the proportions suggested by different authors for the distribution of carbon dioxide in the various reservoirs.

Distribution of fossil fuel produced carbon dioxide into various reservoirs (in per cent) as suggested by different authors.

Reservoir	Bolin and Bischof[8] (1970)	Machta[9] (1972)	Bacastow and Keeling[10] (1973)
Atmosphere	28–42	55	50
Land vegetation	15–20	15	29–36
Ocean	approx. 50	30	14–21

The idea that the vegetation might absorb some of the additional carbon dioxide introduced into the atmosphere and thereby might store the carbon is not too far-fetched. It is known that the plant productivity in glasshouses, where the plants have sufficient light, water and nutrients, can be enhanced when they are grown in an atmosphere containing extra carbon dioxide. It is possible that this same effect prevails outside the glasshouse and is also applicable to well-irrigated and fertilised farmland. But agricultural products form a part of the second reservoir (see page 4) since they are only stored for relatively short times before they are consumed, and their carbon is 'burnt' in respiration to form carbon dioxide again.

According to the models mentioned above, the carbon must be stored for a longer period of time. Thus the calculated distribution can only be realistic if the third reservoir, that is the world's store of wood and humus, is increased in size. In the forests, however, it is usually not the amount of carbon dioxide that is the limiting factor but rather the abundance of light, water and nutrients. This is especially true in tropical forests having reached so-called climax conditions; the forest is capped with a complete canopy of leaves so that all the available light is utilised and consequently the governing factor for further growth is literally finding 'a place in the sun'. Consequently, an increase in the amount of carbon dioxide in the atmosphere would not change the situation there very much. From an ecological point of view it is therefore difficult to see where one could find the vegetation, that, according to the modellers, is supposed to remove or absorb some of this excess carbon dioxide. 'We know of no evidence for increased wood storage in natural forests because of CO_2 increases,' write Woodwell, Whittaker and co-authors in a joint paper on the carbon dioxide problem.[11]

The demands of the meteorologists and geophysicists are, however, rather modest. Assuming that fossil fuel combustion is solely (or at least mainly) responsible for the atmospheric carbon dioxide increase, the quantity of carbon dioxide not accounted for, i.e. carbon dioxide which did not remain in the atmosphere between 1860 and 1970, is about 150 thousand million tonnes. Of this, according to the model, two thirds went into the oceans leaving the rest, about 50 thousand tonnes of carbon dioxide (i.e. 14 thousand tonnes of carbon), to be accommodated by land vegetation. The biomass of the world's forests contain more than 700 thousand million tonnes of carbon and some 1 to 3 billion tonnes are stored in humus. Thus if these two reservoirs for carbon have increased in the last hundred years, perhaps each by 1 per cent (a change that is much smaller than the measurement accuracy in ecology) then this could more than account for the 'missing' carbon dioxide and make the modellers happy. That this really might have been the case cannot be denied by the ecologists even though they know of no mechanisms to perform this task. Thus the problem could be considered as solved, if the forests had been left in their natural state. However, as has been described in part 1 of this book, this is certainly not the case. According to the bioscientists, the biomass and the amount of humus material have been radically decreased by forest clearing. Thus, in addition to fossil fuel burning, large amounts of carbon dioxide have been fed into the atmosphere from this source too.

Is world biomass growing or shrinking?

When we talked together at the Dahlem conference, G. M. Woodwell told me that Lester Machta, the well-known meteorologist, had once asked him casually if he thought that forests might increase their photosynthetic production if the supply of carbon dioxide was increased.

'I replied to him off-hand', Woodwell said, 'that this might happen in certain forests not limited by water or nutrients. This conjecture uttered in a conversation has since been used widely by meteorologists and others involved in modelling the world's carbon budget. I myself have however come to the conclusion that I was wrong then.[12] But even assuming that there are forests which produce more organic substance with increasing atmospheric carbon dioxide, the increases would be of minor importance on a world scale. The models of Machta and others can only be correct if the *total* biomass of the world's forests has increased within the last hundred years. That seems very doubtful to me because changes in the forest area through cutting have had much more influence on the size of the world's biomass, and therefore on the composition of the atmosphere, than any possible increase in photosynthesis. The conclusion must be that the increase in atmospheric carbon dioxide is caused not only by burning fossil fuels but also by liberating carbon dioxide through the destruction of forests and humus.'

The assumption of Machta and other Earth scientists (meteorologists, geophysists, oceanographers etc.) that green plants have stored some of the industrially produced carbon dioxide, can only be true if the total amount of organic matter on the earth's surface, together with the total amount of humus material, has increased. Not only Woodwell disagrees with this proposition, but so do most biological scientists (botanists, ecologists, foresters etc.). On the other hand the world's biomass (including humus) can only have decreased in size if the oceans have absorbed appreciably more carbon dioxide than hitherto assumed or if some other unknown sink for carbon dioxide exists. Both of these possibilities are considered unlikely by most Earth scientists.

The fact that two different models of the world carbon budget exist, each of them recognised by scientists from their own discipline but in contradiction with the model recognised by the scientists from the other discipline, clearly emerged at the Brookhaven symposium on Carbon in the Biosphere held near New York in 1972.[13] The discussion reached its peak at the Dahlem conference in November 1976 and was continued thereafter at many other international interdisciplinary conferences.[14]

Obviously scientists involved in such a dispute start by defending the view of their own discipline and looking for weaknesses in the arguments of their opponents. However, if the opposing point of view is repeatedly stated with convincing arguments, they start wondering whether there might not be weaknesses and unproven assumptions in their own model which may need revision, at least in certain respects.

This point was certainly reached at the IIASA workshop on Carbon Dioxide, Climate and Society in February 1978. At this meeting, though the gap between the two sides was still clearly delineated, there was a definite tendency for *rapprochement* and everyone agreed that only when the basic carbon cycle is properly understood is it possible to make meaningful recommendations about current and future energy policies.

'We now know that we know much less about the carbon cycle than we thought we knew a few years ago,' said Bert Bolin in one of the introductory lectures. He went on to tell the conference that his group in Stockholm is now trying to construct a much more sophisticated model of the carbon cycle than they had used previously, incorporating knowledge from all branches of science and using data that have been checked and cross-checked with all the relevant information that is available, including models of the cycles of other elements, since these of course must be consistent with the model for the carbon cycle.[15]

Bioscientists are always at a disadvantage compared to meteorologists, geologists and other Earth scientists when such a conflict arises because they have no 'hard figures'. The only point that cannot be in doubt is the long-term trend. Three thousand or even one thousand years ago there was certainly more forest than exists today. But what changes have taken place in the last hundred years, and in particular since 1958 when precise measurements of the carbon dioxide concentration started at the Mauna Loa? 'The data are not as good as one might hope,' writes Woodwell, 'but they seen compelling to those who are familiar with them.'[16]

The fact remains however that bioscientists cannot tell with any degree of confidence exactly how large the world's biomass is today. Authors are at wide variance in their assessment of its size. The estimate of Whittaker and Likens, which is usually taken as the most reliable and which is repeatedly quoted in this book (see

tures of the carbon cycle by studying the changes in the relative abundance of different carbon isotopes in the yearly growth rings of trees. As this sort of research uses methods of the 'exact sciences' it has a good chance of being considered 'more respectable' by geophysicists and other Earth scientists than the 'vague estimates' of ecologists and forestry people.

As mentioned before (see page 46) carbon is fractionated in photosynthesis with the lighter isotope carbon-12 being preferred. Hence all carbon compounds of biologic origin – be it from plants, wood, humus or fossil fuel – and carbon dioxide produced by burning, respiration or decay of such biologic carbon contains a smaller concentration of the heavier carbon isotopes than the carbon dioxide in the atmosphere. Moreover radioactive carbon-14 looses most of its radioactivity in some 100 thousand years so that, as was mentioned earlier, fossil fuels, being much older, contain virtually none of this isotope. In wood and humus however, there is only a small change from the ratio in the free atmosphere. Thus there are three reservoirs of carbon that can be distinguished by their relative abundance of the carbon isotopes.

When large amounts of carbon dioxide from either of the other reservoirs enter the atmosphere, the proportion of the carbon isotopes in atmospheric carbon dioxide is changed and such a change will be conserved in the tree-rings of the respective years. One can show from such analysis of tree-rings that the relative abundances of the carbon isotopes changed in the past and thereby with hindsight calculate how much carbon dioxide was released into the atmosphere from the various reservoirs.

In practice all this is not yet as easy or as exact as it sounds. In this new field of research, the scientists are still at the stage of calibrating their instruments. So far one does not yet know for sure with what degree of precision tree-rings reflect changes in the atmosphere; nor can it be explained why different samples taken from the same ring of the same tree – not to mention other trees – often show a considerable scatter in their carbon isotope content. At present, workers in this field are still trying to develop a technique enabling them to extract meaningful information from signals that are so faint that they lie nearly in the range of the background noise. It is therefore easy to understand that various

table 2 on pages 172–3) applies to the position as it was in the 1950s and some of these figures are now well out of date. Of the 17 million square kilometres of tropical rain forest shown in this table, only 9.5 million square kilometres were left in 1975 according to a press release of the World Wildlife Fund.[17] For large areas of the globe, there is no useful information about changes in the afforested areas over the last hundred years, and simply no information at all as to whether the increase in atmospheric carbon dioxide concentration has caused an increase in the growth rate of the forests anywhere in the world.

Using observations from satellites, it is hoped that, in the future, more exact information about changes in the area covered by different forms of vegetation may be obtained. It may then be possible in a few decades to decide whether the northern boundary of the forests is indeed moving northward as Woodwell thinks possible (see page 13).

'Could there possibly be a feedback mechanism operating on the northern forest boundary?' I asked Bert Bolin at the IIASA meeting. 'In a worldwide warming of the atmosphere, caused by an increase in its carbon dioxide content, the northern boundary of the forests might move northward resulting in an increase of area and biomass of boreal forests. More carbon dioxide would be used in photosynthesis, the atmospheric carbon dioxide concentration would decrease and global warming would at least be slowed down.'

'Your reasoning is correct in principle,' was the answer. 'I am however somewhat doubtful as to whether a shift in the climatic zone northwards in one area of the globe could occur without associated shifts of climatic zones elsewhere, that might have the opposite effect on the total biomass reservoir.'

I replied that changes in temperature, as we will see later in this book (see page 97), are much more pronounced in high latitudes of the northern hemisphere than in low latitudes. It seems therefore feasible that shifts in the northern forest boundary could occur without simultaneous shifts in the vegetation zones in low latitudes.

'Well, maybe,' said Bolin, 'but a general warming would speed up the metabolism of bacteria decomposing litter and humus. Carbon dioxide production would thus increase and we could have an opposite effect as well.'

Ulrich Hampicke, who is in the Environment, Energy and Society group at the University of Essen, Federal Republic of Germany, has made an approximate calculation of the relative importance of these two effects. He told me in a letter that he has come to the preliminary conclusion that they are probably about equal in magnitude and consequently cancel each other out.

The worldwide satellite observation programme mentioned above can only provide information on the changes in the *areas* of various types of vegetation. This information is of course important but unfortunately not all that is required. Satellite measurements do not give information on whether the biomass within a certain area is changing and of course they give absolutely no information at all about the amount of humus.

'The size of the humus reservoir has been underestimated for a long time,' the American ecologist W. A. Reiners told me at the Dahlem conference. 'Four years ago at the Brockhaven symposium it was widely suggested that the carbon content of this reservoir was between 700 and 800 thousand million tonnes. At that conference I drew attention to the fact that this carbon reservoir might be ten times as large if Delwiche's estimate of the amounts of nitrogen in the soil are anywhere near correct.'[18]

'You mean that since organic substances have a relatively fixed proportion of carbon to nitrogen content, one must assume that this proportion also holds for humus?'

'Yes, exactly, and because of the wide divergence between these estimates, I suggested that new measurements be made to establish more precisely the size of the humus reservoir. Several scientists have actually been working on this problem since then and their results seem to me to be far more acceptable. Here in my paper, I referred to results of a study made by W. H. Schlesinger, that is currently in print, where he estimates the humus reservoir to contain 1.5 billion (1.5 million million) tonnes of carbon. Bolin quoted a paper by H. L. Bohn suggesting that the carbon in this reservoir could well be around the 3 billion tonnes figure.'[19]

'And how do man-made changes in vegetation affect humus?' I asked.

'Generally speaking, in a very bad way. The change from forest

to field means that the soil is ploughed, aerated, and oxidation increases. With the removal of the forest cover, the soil temperature rises, increasing the respiration rate of the microorganisms in the soil. The agricultural products are harvested and distributed, in general, far away from the area in which they were produced and consequently the organic substance is not returned to the soil. If no legumes are planted, then the nitrogen content in the soil can be reduced by some 50 per cent, and since nitrogen is usually in some organic compound containing carbon too, the amount of carbon therefore decreases in the same way. The same thing also occurs when swamps and moors are drained and transformed into arable land. It is possible, with careful management, to maintain or even slightly increase the humus content when forests are transformed into pasture. Usually erosion drastically increases after forest clearing, humus material is removed by the action of wind and water and by this process fertile land can be gradually converted into desert.'

'You therefore think that the amount of humus in the world is at present decreasing?'

'Without doubt. I would even go further to suggest that there is a direct relationship between the world's population growth rate and the decrease in the amount of humus.'

However, humus 'lost' in erosion processes need not necessarily be oxidised. In a conversation at the IIASA workshop, the ecologist, Professor Helmut Lieth, who, after a number of years in the USA has now returned to his native Germany, referred to the 'black' or brown rivers of Brazil, Siberia, and elsewhere. 'The colouring is caused by humus,' he told me, 'and if the forests in the catchment areas of such rivers are cut then they will presumably transport even more such dark material. I would think however that a considerable proportion of this humus reaches the sea in an unoxidised state, and is deposited there as *reduced* carbon in the ocean sediments. Consequently, not all humus losses need increase the carbon dioxide burden of the atmosphere.'[20]

What can be learnt from tree-rings?

In recent times bioscientists got unexpected support from a newly emerging branch of research which attempts to analyse past fea-

groups dealing with these low signal levels are producing results that are at some variance with one another. It was said at the IIASA meeting in 1978 that it was hoped that these 'teething troubles' would be overcome in the next five to ten years and then, alongside the estimates of the ecologists, there might be an independent and reliable method to provide information about the size of the biological source of carbon dioxide.

One thing seems to be clear, however, from the tree-ring analysis made so far. At least for the first part of this century, when consumption of fossil fuels was far less than it is today, the net influx of carbon dioxide into the atmosphere from this source was considerably less than that from the biosphere. According to the American biologist and geologist Minze Stuiver, the net biospheric source of carbon dioxide has however decreased since about 1930 and is now considerably smaller than it was at the beginning of this

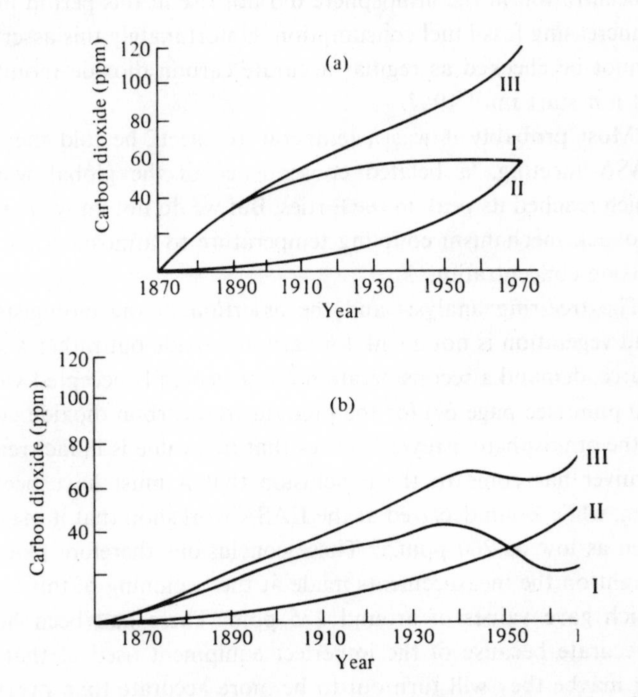

Figure 10 Cumulative amount of carbon dioxide released into the atmosphere from non-fossil sources (I), fossil sources (II) and all sources combined (III) according to (a) Stuiver and (b) Freyer.

century or even no longer existent.[21] That would mean that the destruction of the tropical rain forests in the last few decades, which no one can dispute, has been largely or even completely compensated by afforestation and an increase in the growth rate and the area of forests in other places.

H. D. Freyer of the Nuclear Research Centre in Jülich, West Germany, presented at the IIASA meeting a more detailed picture of the development since 1930.[22] His tree-ring analysis showed that until the late 1930s the biosphere acted as a source of carbon dioxide and then between about 1940 and 1960 it acted as a *sink*. Since then it has reversed its role again to that of a source liberating now not much less carbon dioxide into the atmosphere than is released by the combustion of fossil fuel.

According to Freyer's analysis land vegetation was such an effective sink between 1940 and 1960 that *total* carbon dioxide concentration in the atmosphere did not rise in this period in spite of increasing fossil fuel consumption. Unfortunately this assertation cannot be checked as regular accurate carbon dioxide monitoring did not start until 1957.

'Most probably it was a temperature effect,' he told me at the IIASA meeting, 'a belated consequence of the global warming which reached its peak in the forties. But we do not know as yet the feedback mechanism coupling temperature to atmospheric carbon dioxide concentration.'

The tree-ring analysis and the assertion of the biologists that land vegetation is not a sink for carbon dioxide but rather a strong source, demand a reconsideration of the generally accepted value of 290 ppm (see page 67) for the preindustrial carbon dioxide content of the atmosphere. Freyer believes that this value is in fact realistic, Struiver has come to the conclusion that it must have been 268 ppm, while Bolin declared at the IIASA workshop that it may have been as low as 260 ppm.[23] These conclusions therefore place new weight on the measurements made at the beginning of this century which gave values of around 275 ppm. These had been held as inaccurate because of the imperfect equipment used at that time, but maybe they will turn out to be more accurate than previously believed.[24]

According to evidence submitted to the Dahlem conference by George Woodwell, the annual clearing rate of tropical forests is about 1 to 2 per cent of their 1950 area.[25] Other sources (some of them quoted on pages 10–12 of this book) give similar estimates. Assuming an average world rate of wood harvest of only 1 per cent, of which about two thirds are transformed into carbon dioxide within a short period of time (the rest remaining stored as lumber), then the annual carbon dioxide release would be about 18 thousand million tonnes (5 thousand million tonnes of carbon). This is about the same amount as produced by burning of fossil fuel.

'Don't you think that two thirds might be too low an estimate, when one allows for slash-and-burn agriculture, the amount consumed as wood fuel, and the production of newsprint and other short-lived articles made from wood?' I asked Woodwell.

He raised his brows and looked at me in amazement. All through this conference he had been harassed that his estimates were *too high*. Now somebody asked him whether they might be too low!

'You are probably right,' he replied, 'but I was trying to submit a very conservative estimate.'

In his effort to be on the conservative side, Woodwell also had made rather generous assumptions about the amount of carbon stored in secondary vegetation after the destruction of primary forests, while making a rather low estimate for the amount of humus oxidised when forests are cut. He therefore came to the conclusion that the size of the biospheric carbon dioxide source is most probably within the range of 15 to 30 thousand million tonnes of carbon dioxide (4 to 8 thousand million tonnes of carbon) per year.

'It might easily be considerably more,' he added, 'but it is very unlikely to be much less.'

Making a liberal rather than a conservative estimate, it might in his view be as much as 66 thousand million tonnes of carbon dioxide (18 thousand million tonnes of carbon) a year. That would be three and a half times as much as from the fossil fuel source.[26] Bill Reiners told me at the Dahlem conference that he considered

Woodwell's maximum estimates to be too high. Nevertheless he later co-authored a paper of Woodwell, Whittaker and others, published in *Science* (see reference 11), which presented a wide range of possible values, for he is in agreement with the main line of the argument, that the biosphere is not a sink for carbon dioxide but rather a source of considerable importance.

One can understand the sighs of the Earth scientists when they are confronted with such a diverse range of estimates. How can they include realistic figures into a model calculation when the bioscientists themselves cannot agree on a common value? The fact remains, however, that even the minimum estimate made by the biologists for the size of the biospheric source of carbon dioxide is not much smaller than the amount of carbon dioxide released by the combustion of fossil fuels. Thus when one takes account of all sources, considerably more carbon dioxide is released into the atmosphere than has been considered in the model calculations so far. Yet the amount in the atmosphere is not going to change just to please the model builders. As a matter of fact it is one of the very few undisputed figures in the whole game.

The accurate measurements made over the last twenty years show a steady increase which is running at present at the rate of about 8 to 10 thousand million tonnes of carbon dioxide (2.2 to 2.7 thousand million tonnes of carbon) per year. The question therefore arises: Where does the rest go to? Even the most conservative estimates of the bioscientists suggest that the rest is not 10 thousand million tonnes, as suggested by the Earth scientists, but rather of the order of 25 thousand million tonnes of carbon dioxide (7 thousand million tonnes of carbon). The oceanographers present at the Dahlem conference declared that the oceans could not possibly absorb more than about one third of this amount. And so one started searching for some other sink for carbon dioxide.

One suggestion was that there might be a stronger flow of carbon into the sediments, i.e. what we called the fourth reservoir. However, the annual rate of sedimentation for carbon of biological origin is, according to Schidlowski and co-authors (see page 62), at most 150 million tonnes (corresponding to about 500 million tonnes of carbon dioxide) and Garrels and co-workers give an even lower value. It is very unlikely that this rate would increase even if more carbon dioxide was available, for in the oceans it is not the amount of carbon dioxide but rather the short supply of nitrogen and phosphorus that limits the growth of algae.

The Canadian oceanographer C. S. Wong later proposed that the amount of organic carbon buried by sedimentation in *coastal regions* of the sea may have been underestimated. In particular, he suggested that the Arctic waters, 'where a short period of production is combined with a large input of sedimentary material from the continents, may be favourable to this concept', all the more so since decay is slow in cold polar waters. Making very generous assumptions he comes to the conclusion that the annual worldwide sedimentation rate could be 1.2 thousand million tonnes of carbon (equivalent to 4.4 thousand million tonnes of carbon dioxide).[27] However, this still only accounts for perhaps one quarter of the shortfall and suggests that some other sink must still be found.

Is it not possible, the participants at the Dahlem conference questioned further, that excess fertilisation (eutrophication) of fresh-water lakes observed on many occasions might provide a sink for the 'excess' carbon? Rough calculations show that this effect would accommodate only a few per cent of the sink required.

Thus, various hypotheses have been advanced, and as Woodwell, Whittaker and co-workers summarised in their paper published in *Science*, 'few of them can be ruled out unequivocally, but there appears to be no single major sink for carbon, equivalent to the amount estimated to be released, except in the oceans'.[28] On the other hand, if the biosphere has in fact been a source of carbon dioxide rather than a sink, and if there is no other major sink, one would require a model where the oceans accommodate about five times as much carbon than hitherto assumed. 'It is difficult to see how the previous estimates of the oceanic uptake of carbon dioxide could be in error by a factor of five,' declared A. Lerman as one of the group rapporteurs at the Dahlem conference.

Christian Junge, the well-known German expert on air chemistry, did not agree. 'The oceans are a very big reservoir,' he said in the closing session of the conference. 'They contain about sixty times as much carbon dioxide as the atmosphere. If the mixing rate of surface and deep waters is only a little bit higher than assumed, then, as far as I can see, all the discrepancies would be resolved. As yet we are still a long way from being able to define worldwide circulation and gas exchange patterns to such a degree of confidence that we could rule out the possibility of a much bigger carbon dioxide uptake by the oceans than has been postulated by the models used up until now.'

At the IIASA workshop in February 1978, Wallace Broecker, as the spokesman for the oceanographers, took a much more conciliatory position than he had taken at the Dahlem conference. He said that, to a first approximation, one does not need a biospheric carbon dioxide sink and that even a small biospheric source might be possible, though not nearly as big a source as Woodwell and his colleagues assume.

'Can you imagine any other major carbon dioxide sink that as yet has not been considered?' I asked him during a break at the workshop.

'Well, personally I doubt it,' was his answer. 'But Garrels recently suggested that magnesium calcites in the oceans might be such a sink.'

So I wrote to Professor Garrels enquiring about this. He replied: 'With regard to oceanic sinks, I have suggested the possibility that magnesium calcites may be approximately in equilibrium in surface sea water. If so, they would dissolve with increasing carbon dioxide, even though calcite and aragonite could not because the surface seawater is supersaturated with respect to them. Whereas this is a sink in principle it has neither been adequately tested nor assessed quantitatively. Another possible sink is the dissolution of calcites with low magnesium content and aragonites in high latitude waters. Alexandersson of Sweden has shown that 100 000 square kilometres of the Skagerak-Kattegat have such aggressive surface waters.'[29]

A further terrestrial sink was suggested at the IIASA meeting by Karl Erik Zimen, a Swedish nuclear chemist who had been director of the Hahn-Meitner-Institute for Nuclear Research in Berlin for many years.[30] He suggested that part of the carbon dioxide produced by the oxidation of humus is not released into the atmosphere but rather immediately dissolved in subsoil waters and thus becomes part of the carbonate cycle. Consequently, the burden that the atmosphere would need to bear due to the destruction of humus, assessed by Woodwell and co-workers to amount to about 0.5 to 5 thousand million tonnes of carbon annually (some 1.8 to 18 thousand million tonnes of carbon dioxide) could be considerably less (see also page 80). However, a panel on Energy and Climate set up by the American National Academy of Science independently checked this possible sink and came to the conclusion that it is unlikely to be of any real significance.[31]

Last but not least a new suggestion was submitted by Woodwell in a paper for the IIASA meeting. He thinks that faecal pellets from marine animals falling through the water at rates considerably higher than plankton could provide the mechanisms for a speedy transport of *considerable amounts* of carbon from the surface waters to the deep waters and to the bottom of the sea. 'The data in support of such a hypothesis are limited,' he admits, 'but the broad pattern is reasonable. The role of the biota in oceanic transfers may be larger than previously assumed.'[32]

So far all these ideas are mere suggestions that need to be checked and quantified. The problem of the missing sink has not yet been solved. However, after several years of intensive discussions one can expect there will be at least some common ground on which the bioscientists and Earth scientists can meet.

Forecast with many unknowns

Differences of opinion as displayed between bioscientists and Earth scientists on the carbon cycle can give a new impetus to research and are in general very helpful for the progress of science. However, for those involved in trying to predict the development of the carbon dioxide concentration in the atmosphere, these differing opinions make this task nearly impossible. If it is true, as the bioscientists suggest, that during the last hundred years considerably more carbon dioxide has been released through the destruction of forests and the humus layer than has been absorbed by vegetation and thus the total biomass has been reduced in size rather than increased, then the model calculations of the Earth scientists are wrong on *two* counts. Firstly because they base their calculations, and hence their mode of distribution, on the erroneous assumption that the net absorbtion of carbon dioxide by the biosphere exceeds the amount that it releases. Secondly because the figures used in the calculations for the amount of carbon dioxide released, mainly through the combustion of fossil fuels, are far too small and the release of carbon dioxide from the destruction of forests and the humus layer is not taken into consideration. There is more carbon dioxide to be accommodated.

Conclusions drawn from such a questionable basis must be considered with a good deal of caution. However, since the models incorporate *two* mistakes that deviate in opposite direction it is possible that at least to some extent these errors cancel each other. After all, these models have been tailored in such a way that they can correctly 'predict' the increase in the carbon dioxide concentration of the atmosphere that has been observed so far. Consequently, it is likely that the carbon dioxide increase predicted by the models for the next twenty to thirty years is pretty near the mark despite the possibility that the calculations contain some errors. Naturally, the further one looks into the future, the greater the effect these errors will have.

Moreover it is not only the methods of calculation which are questionable but some of the input data as well, since the carbon dioxide prognosis must be based on other forecasts which are themselves disputed. How will the consumption of fossil fuels develop over the next few decades? Will the destruction of forests and the humus layer be even further accelerated by pressure from an increasing population? Will there be an appreciable amount of reforestation? Question after question can be raised and one can only make a guess at the answers. Perhaps it would be the most 'respectable' thing to do to declare that atmospheric carbon dioxide concentration will increase, but that the volume and the growth rate of this increase cannot be predicted as yet with any degree of confidence.

Such a declaration however would be most unsatisfactory for an increase in atmospheric carbon dioxide above a particular threshold level (the size of which is disputed) would probably produce grave changes in the climate. Therefore one simply *must* know in advance what will happen if a certain set of circumstances prevails. The problem needs to be solved, despite the large number of unknowns, at least to such an extent that one can say what will happen to one of the unknowns if one makes certain assumptions about the others.

At the Brookhaven symposium in May 1972, Bacastow and Keeling presented their predictions for the increase in atmospheric carbon dioxide over the next hundred years based on a 4 per cent growth rate in the use of fossil fuels. Only one and a half years later however these figures proved to be unrealistic. Following the oil crisis of 1973, the consumption of fossil fuels did not grow at all in 1974 and 1975 and economists predict that future growth rates will remain below the pre-1973 trend. A representative of the OECD Secretariat in Paris gave three reasons for this at the Dahlem conference. The general slowing in the economic growth rate, the strong incentive to conserve energy resulting from the increased price of oil and the growing share of nuclear energy in total energy production.[33]

'If you look at Keeling's diagram for the development of the release of industrial carbon dioxide,' C. E. Junge told me, 'you can see that there was a period between about 1910 and 1945 when there was a very irregular and slower growth rate (see page 70). How do we know that we are not at the beginning of such a period today?'

This is perhaps not a very comforting view, but detailed questioning reveals that most economists agree that we really don't know. The various economic institutes tend to change their trend predictions every few months and so in principle everything seems possible except a return to the relatively undisturbed development that took place before 1973.

There is, however, another way of looking at this problem. The long-term predictions for the world population growth give a figure of about 2 per cent per year and one would expect that energy production *should* grow at an even faster rate than this, especially in developing countries where population grows fastest and misery is greatest. Seen in this light the 4 per cent growth rate for worldwide energy consumption suggested by Bacastow and Keeling as a long-term average would not seem to be an exaggeration. But such a rate, of course, does not take account of whether there are indeed sufficient supplies of fossil fuels to meet this demand.

'An exponential growth rate is difficult to imagine considering the limited reserves of fossil fuels,' stated K. E. Zimen, who is also working on carbon dioxide forecasts. Indeed, present predictions suggest that the reserves of natural gas and oil will be virtually exhausted in the first decades of next century. However, the reserves of coal and oil-bearing shales are so large that one could envisage a strong growth rate being maintained until the end of the twenty-first century. There are also technological developments that now enable highly favoured liquid and gaseous fuels to be extracted from these raw materials.

From this point of view, the study by Bacastow and Keeling probably does not provide a very realistic forecast, but one can use it for what the futurologists call a 'scenario': a study of what would happen if ... In such a scenario it could be checked what would be the consequences if one tried to supply the energy demands of a world population of ten to twelve thousand million people mainly with fossil fuels up to the middle of next century. The forecast of Bacastow and Keeling shows that this simply is not possible. With such an increase in the consumption of fossil fuels during the next hundred years, the amount of carbon dioxide released into the atmosphere would amount to about eight times that contained in the atmosphere today. Even if one were to take the optimistic view that only about one half of this amount would remain in the atmosphere, this would still result in an increase in the carbon dioxide concentration by a factor of about three or four compared to today's levels. This would most probably lead to climatic consequences that are truly unacceptable. Bacastow and Keeling are of the opinion that the situation could be even worse as the oceans might not be able to keep up their absorption rate because of their changed chemical composition. Consequently, they predict that such a set of circumstances would result in an atmospheric carbon dioxide concentration of six times today's level.

Zimen has presented calculations[34] for the situation where the fossil fuel consumption in the next hundred years levels off. Such a situation encompasses the widely held view that nuclear energy, and perhaps other forms of energy production as well, will contribute an increasingly greater part of the total energy budget. However, the results of these predictions, albeit more realistic than those of Bacastow and Keeling, are no more consoling. The carbon dioxide content of the atmosphere would, in the year 2000, reach a level of about 130 to 140 per cent of the 1860 level and, with the further development in the use of fuels, a doubling could be expected within the next fifty to hundred years. Only if the average long-term growth rate of fuel consumption drops to about 1 to 1.5 per cent per annum, will the atmospheric carbon dioxide concentration remain under this doubled level for the next hundred years. 'We can utilise all the fossil fuel reserves that exist', comments Zimen, 'as long as we don't do it too quickly.'

Zimen and co-workers calculated atmospheric carbon dioxide concentrations for an increasing, a constant and a decreasing world biomass, and according to their model this would have virtually no influence on the carbon dioxide forecast for the next thirty to fifty years. In their opinion, it is the way in which fuel consumption develops that holds the key. However, they only assumed a relatively small deforestation rate, releasing about 20 per cent of the

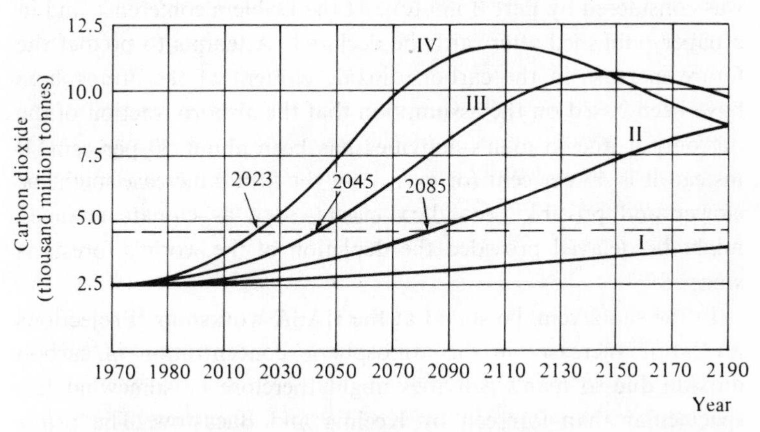

Figure 11 Projected increase in atmospheric carbon dioxide concentration for annual growth rates of 1 per cent (I), 2 per cent (II), 3 per cent (III) and 4 per cent (IV) of fossil fuel consumption based on the assumption that the biological cycle is balanced (i.e. there is no net consumption or net production of carbon dioxide by the biosphere). For large growth rates, higher maximum values of atmospheric carbon dioxide are reached, but the reserves of fossil fuel are exhausted much earlier as compared to small growth rates. The dashed line indicates the level of doubling the present carbon dioxide amount; the corresponding figures give the respective year when this level will be reached. (From K. E. Zimen *et al.* in: Stumm (ed.), p. 296.)

amount of carbon dioxide produced by the combustion of fossil fuels. A forest destruction rate of the size suggested by Woodwell and others cannot be accommodated in Zimen's model with the Mauna Loa data of the past evolution of atmospheric carbon dioxide concentration, unless assumptions about the carbon dioxide absorption of the oceans are made which oceanographers consider to be entirely unrealistic. On the other hand, even to accommodate a relatively small deforestation rate, Zimen had to assume a very high 'biota growth factor' for the remaining vegetation, i.e. he had to assume that plant growth is enhanced considerably by even a rather small increase in atmospheric carbon dioxide concentration. This, however, is an assumption that is held to be highly unlikely by the bioscientists. Thus the mathematical description of this problem emphasises once more that a model acceptable to both Earth scientists and bioscientists cannot be constructed unless an additional carbon dioxide sink is found.[35]

The question of how the carbon dioxide predictions might change if the arguments of the biologists are taken into account, was considered by Bert Bolin too. At the Dahlem conference and in a paper published afterwards he declared: 'Attempts to predict the future increase in the carbon dioxide content of the atmosphere have been based on the assumption that the airborn fraction of the net output due to man's activities has been about 50 per cent. If instead it is 40 per cent (or even less), the future increase might be slower and possible secondary effects, such as climate changes, might be delayed provided the depletion of the world's forests is stopped.'[36]

In the same vein, he stated at the IIASA workshop: 'Projections for future increases in the atmospheric concentration of carbon dioxide due to man's activities might therefore be somewhat less spectacular than foreseen by Keeling and Bacastow. The prime reason for this tentative conclusion is that the oceans and the bottom sediments might serve as more effective sinks than has so far seemed possible.'

However Bolin combined this slightly optimistic note with a word of warning: 'The continued burning of fossil fuels until the presently known reserves are exhausted will lead to a manifold increase in the atmospheric carbon dioxide concentration; this

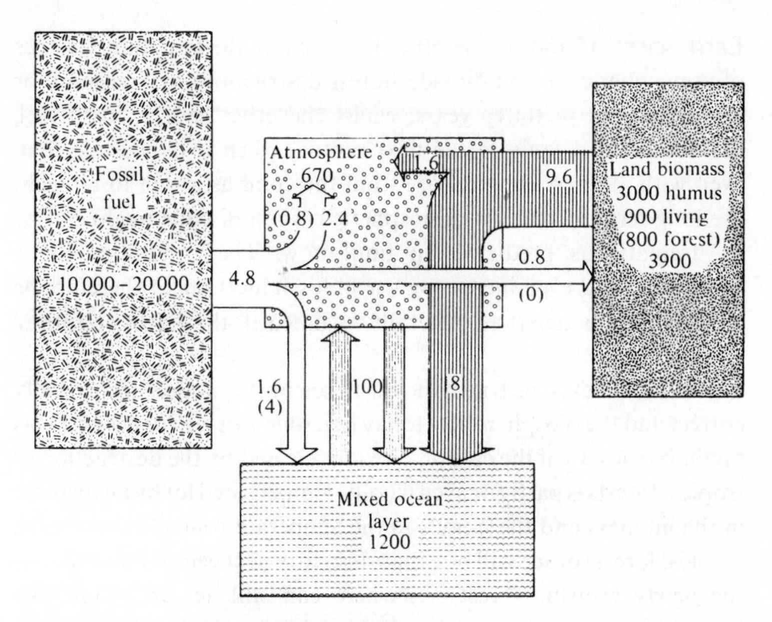

Figure 12 Schematic representation of the modified global carbon dioxide cycle. Original model of geoscientists is represented by white arrows. Hatched arrows represent biogenetic flux hypothesis. The subsequent change in fossil fuel carbon dioxide fluxes is given in parentheses. (From A. Nir *et al.* in: Stumm (ed.).)

expectation certainly warrants serious consideration with regard to possible climatic changes.'[37]

In any long-term prediction it is also important to remember that the carbon reservoir of the forests and the humus layer is considerably smaller than that of the fossil fuels. If the destruction of the tropical rain forest is really as drastic as Woodwell, Adams and Brünig suggest, then this process could well be completed within the next thirty years. This would mean that there would be virtually no tropical rain forest left that could be cut afterwards. Provided that the boreal forest of Siberia and Canada are not then subjected to the same sort of wanton destruction that we observe in the tropics today, it is possible that the world biomass might then no longer decrease but even increase.

But if on one hand the sinks for carbon dioxide (mainly the oceans) absorb considerably more than hitherto assumed by the

Earth scientists and on the other hand one of the two main sources of atmospheric carbon dioxide (forest destruction) will be more or less exhausted in thirty years, whilst the other source, fossil fuel, reaches its peak only in the next century, then it is possible that, even with an increasing release from the combustion of fossil fuels, the long-term increase in atmospheric carbon dioxide might not be as dramatic as predicted up until now. The conditions for a continuing and even increasing use of fossil fuels might therefore be considerably better than generally assumed in the Earth scientists' models.[38]

This would even be true if the Earth scientists' models are virtually correct and the biosphere acts today *as a source as well as a sink*. (This might be the case if the carbon dioxide released by the destruction of tropical forests is partly or even wholly compensated for by an increase in the biomass and the humus in the temperate zone.) For while the tropical forest source will be exhausted early next century in any case, the newly growing forests elsewhere can still act as a sink. No bioscientist would disagree that the biosphere *could* act as a very large and effective carbon dioxide sink if there were a large-scale worldwide reforestation. (More about this in part 5.)

Danger for the climate

Carbon dioxide, as has already been mentioned in this book, while being a good transmitter of sunlight, demands a heavy 'customs duty' from the infrared radiation going back from the Earth into space (see page 35). This produces the so-called greenhouse effect that prevents a drastic cooling of the Earth during the night. Increasing the amount of carbon dioxide in the atmosphere reinforces this effect and is expected to result in a warming of the atmosphere near the Earth's surface.

Up to now such an effect has *not* been observed but this should not be taken as a basis for believing that increasing the concentration of carbon dioxide in the atmosphere has no effect on atmospheric temperatures; rather the effect is camouflaged so far by larger natural fluctuations. To determine the trend in such fluctuations disregarding short-term 'accidental' changes, five-year running means have been used (for example the years 1966–70, 1967–

71, 1968–72, etc.). Such a method of calculation suggests that the amplitude of oscillation for the northern hemisphere taken as a whole was less than one degree centigrade in the last hundred years. Considerably larger oscillation of the order of 2.5 degrees were observed in the Arctic region. The trend displayed in the two curves (figure 13) is consistent and shows a lower than average temperature for the period 1880 to 1920, an above average temperature for the period 1920 to 1960 and a cooling during the sixties followed by a further warming.[39]

The cause of these natural oscillations is not as yet fully understood but they are certainly the result of the interaction of several independent variables: small variations in the intensity of solar radiation, small changes in the orbit of the Earth and the tilt of the Earth's axis of rotation, differing amounts of dust released into the

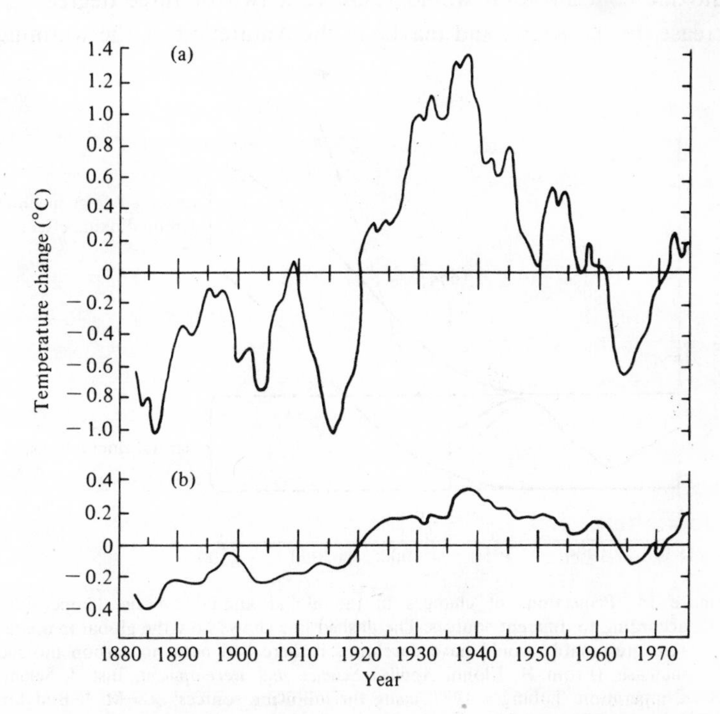

Figure 13 The long-term trend of the air temperature oscillations in the northern hemisphere (five-year running averages). (a) The latitudinal belt 72.5°–87.5° N. (b) The latitudinal belt 17.5°–87.5° N. (From M. I. Budyko and K. Y. Vinnikov in: Stumm (ed.).)

stratosphere through volcanic eruptions changing the transmission characteristics of the atmosphere for sunlight, etc.

It is likely that all the factors that influence the Earth's climate are not yet known and it is certain that the way these effects are amplified or quenched by interacting feedback mechanisms is not yet understood. Consequently, forecasts made today about future trends in the climate are not free from large uncertainties. One must rely in many cases on a 'hunch' or a simplified hypothesis rather than on a sound theoretical basis.

Under such circumstances it is possible only to make rather approximate predictions about the effect that increasing the carbon dioxide concentration would have. It is suggested that by the end of this century the global average annual temperature will have risen by about half a degree centigrade and that doubling the carbon dioxide concentration would result in a two or three degrees increase. In the Arctic and maybe in the Antarctic too, the warming

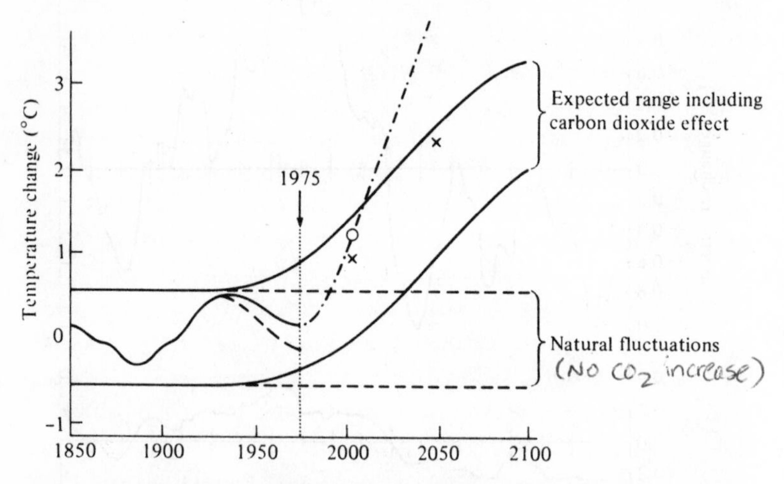

Figure 14 Projections of changes in the global annual average temperature according to different sources. The dashed line shows how the global temperature presumably would have developed if there had been no carbon dioxide increase. (From H. Flohn, *Applied Science and Development*, Inst. f. Scient. Cooperation, Tübingen 1977, using the following sources: × = M. I. Budyko, *Izmenija Klimata*, Gidrometeoizdat, Leningrad, 1974, p. 260; —·—·— = W. W. Kellogg, *Effects of Human Activities on Global Climate*, WMO Report, October 1976; 'Expected range' = J. M. Mitchell, Jr., *Environmental Data Service Magazine*, March 1977.)

could be three or four times as large as the worldwide average change.

To the layman such a change may seem insignificant, since the range of temperatures experienced during the year is far larger than this. However, one cannot use the seasonal temperature variations as a basis for comparing the significance or otherwise to changes in yearly average values. One must therefore find some other basis for comparison. For example, a lowering of the average value by only five degrees could result in the onset of a new ice age while raising the average temperature by two to three degrees would be more than the warming that afflicted Switzerland in the late middle ages and brought a locust plague seventeen times. Such an increase would even be slightly more than that experienced some thousand years ago, when the Vikings sailed on ice-free seas and a country that today is virtually an ice-covered island had green shores. Greenland got its name, seemingly incongruous today, in that time.

Atmospheric carbon dioxide is at the moment the strongest but unfortunately not the only factor resulting from human activities, causing a general warming. The previously widely held view that increasing the amount of dust in the air would cause a cooling of the atmosphere is today rejected by meteorologists. Only dust that penetrates into the stratosphere is now considered to cause this effect. Dust in the air near the ground not only restricts the amount of sunlight reaching the ground but also limits reradiation into space. Depending on the size, colour, chemical composition etc. of these dust particles, various parts of the radiation spectrum are absorbed and warm these dust particles and they in turn warm the air. In many cases the amount of radiation penetrating this heat-protective coating is reduced in both directions, and hence, although there is less solar radiation reaching the ground, there is still a net warming of the lower atmosphere.[40]

An increasing consumption of energy also results in a warming of our environment as is clearly measureable in towns and industrial conglomerates today. Such regions are effectively warm islands in a colder sea since they are a few degrees warmer than their surroundings. Seen on a world-scale, however, the amount of energy released through human activities is still very small compared with the incident energy from the Sun.

Finally, the rising atmospheric concentration of various trace gases also can enhance a worldwide warming. In particular, the increase in the use of nitrogen fertilisers causes an increase in the amount of laughing gas (N_2O) released into the atmosphere. Similarly, the use of spray cans has led to a considerable release of chlorofluoromethanes (the carrier gas used in spray cans). The special chemical properties of these gases and their effect on atmospheric ozone will be discussed in some detail in part 4 of this book. Here it is sufficient so say that they also play a role similar to that of carbon dioxide in the greenhouse effect as they also absorb energy in the infrared (i.e. demand particular customs dues from the radiation passing through the atmosphere) generally in regions not strongly absorbed by other gases, and this absorbed energy results in warming the lower atmosphere. Doubling the amount of laughing gas in the atmosphere, something that is certainly possible within the next fifty years, could result in a global temperature increase of half a degree. The increase in the amount of chlorofluoromethanes might result in a global warming of the order of one degree centigrade.[41] These increases are of course in addition to those caused by carbon dioxide.

General temperature fluctuations, as has been mentioned, are much stronger in Arctic regions than in middle or low latitudes and thus all the factors causing a warming of the atmosphere result in reducing the temperature gradient between the equator and the polar region and thus a weakening of the atmospheric circulation in the northern hemisphere. In a paper submitted to the Dahlem conference, the well-known Soviet climatologist M. I. Budyko suggested that these effects probably would lead to a severe reduction in the annual rainfall in the interior of continents.[42]

In certain areas, particularly near the polar circle, a warming of the order of two degrees might well be considered as a positive effect, but a decrease in rainfall to occur in the much larger and more densely populated regions of the middle and low latitudes would mean a very severe overall deterioration in conditions. Yet even where there is an improvement, the size and extent of which cannot be predicted with any accuracy, it may be a doubtful blessing. For every society, as Budyko underlined in his conference paper, adapts to the prevailing climatic conditions. Consequently *every* change, even one that might be considered an improvement, requires a long-term complicated adaption process, new investment and so on.

A global warming of two to three degrees, something that might well be expected from doubling the amount of carbon dioxide in the atmosphere, would lead to 'unacceptable climatic consequences' declared the German climatologist Hermann Flohn at the Dahlem conference. Such a change, he suggested, would cause a northward shift of the subtropical high-pressure belt by some hundred kilometres. This would possibly lead to a severe reduction in the amount of winter rainfall in California, the Mediterranean countries, the Near East, Pakistan and the Central Asian part of the Soviet Union, and this winter rainfall is essential for maintaining water supplies in these regions.[43]

Other scientists are a little more cautious in making detailed predictions about what might happen. The American meteorologist Lester Machta stated at the Brookhaven symposium that, as a consequence of the increase in carbon dioxide and the associated warming process, various feedback mechanisms would be started, but without a lot of further research he felt unable at that stage to make any predictions about the possible climatic consequences.[44]

Professor C. E. Junge takes a similar view. 'We know far too little at present to be able to make an exact prediction about what effect an increase in carbon dioxide will have,' he told me during a break at the Dahlem conference. 'We haven't really come to grips with these feedback mechanisms and there are as yet no realistic models for them. I too consider it very likely that carbon dioxide will have a considerable impact on the climate, but at present I am not prepared to say what precise effect a certain percentage increase in the carbon dioxide content would have. The effects might be larger than is currently thought likely while on the other hand they might well be smaller.'

'Do you agree with Professor Flohn's suggestion that the expected increase in carbon dioxide will have "unacceptable climatic consequences"?' I asked.

'I agree with Professor Flohn insofar as I would say that carbon dioxide is going to be a serious problem in the next hundred years. Precisely how serious cannot be stated yet.'

At the IIASA workshop in 1978, the climatologists were even more reluctant to make any detailed predictions than they were at the Dahlem conference. 'Temperature is only one measure of the environment in which we live,' said the American climatologist William W. Kellogg in his introductory address. 'Wind patterns and *regional* rainfall trends will be of equal or even greater importance to agriculture, water resources, energy demands and so on. We would dearly like to be able to specify the changes in atmospheric circulation patterns that would accompany a global warming. However, our general circulation models are still too primitive to reproduce them reliably. In particular, the hydrological cycle that determines whether rain or snow will fall is difficult to include properly.'[45]

Similarly, the report of the climatological working group of the workshop states: 'Some regional changes will be much greater than the global average, and they will be both positive and negative with regards to temperature and precipitation. Climatic change will bring with it both gains and losses, winners and losers. Unfortunately, climatologists still cannot say just *where* these will be.'[46]

Certain clues as to how the climate in particular regions would change as the result of a worldwide warming might be found by studying the climatic conditions that prevailed in the Earth's past. After the last ice age there was a period between about eight thousand and four thousand years ago when the average global temperature was one to two degrees higher than it is today. This period therefore had an associated higher evaporation rate and hence more precipitation when considered on a global average. Many (but still not all!) arid areas of today had more rain then.

If you describe the situation existing during that period to representatives from developing countries, particularly the Near East, they are united in their call to return to that past as quickly as possible, Herman Flohn told the IIASA workshop. However, he continued, five thousand years ago Labrador was still covered by glaciers remaining from the last ice age and hence the atmospheric circulation patterns of that time were certainly different from those that would be expected as a result of a global warming today. According to William Kellogg, the causes for the warming then were probably changes in the Earth's orbit around the Sun and not due to a change in the infrared absorption characteristics of the atmosphere. 'Even though the Earth may have been equally warm, the way in which the extra heat was supplied must have been different. Consequently we would not expect circulation patterns to be just the same.'[47]

Nevertheless one might still argue that the risks associated with doubling the amount of carbon dioxide in the atmosphere are not unacceptably large[48] – all the more so since the alternative to increasing atmospheric carbon dioxide through fossil fuel combustion would probably be an increased use of nuclear energy which is fraught with risks as well (more about that in part 5). Thus I asked Herman Flohn in a break of the IIASA workshop:

'What you said today doesn't sound as pessimistic as what you said at the Dahlem conference. Do you now feel that it is acceptable to take the calculated risk of a limited global warming?'

'The thing that worries me the most', came the reply, 'is that neither the size of the warming nor the associated risk is known with sufficient accuracy. If the global warming, through the increase in the amount of carbon dioxide and other trace gases, were four or five degrees rather than two to three, which would mean a warming of ten to fifteen degrees in the Arctic region, then we can expect all the drifting ice covering the Arctic Ocean to melt. This would almost certainly result in an *irreversible* change to a state which has not existed for more than two million years, and no one is in a position to say what the effects might be on a more localised scale.'

According to geological evidence, Flohn continued, the Arctic Ocean was free of ice last time in the late tertiary, i.e. 12 to 2.5 million years ago. At that time parts of Central Europe were arid steppes.

'I would like to make it clear', he added, 'that I expect really *unacceptable* climatic consequences only in the event of an ice-free Arctic Ocean which presumably would be induced by an increase in the atmospheric carbon dioxide concentration to about 2.5 to 3 times the present value. Or would you consider it acceptable to have salt lakes as in Northern Africa in the vicinity of Vienna?'

The reason for this uneasiness about the Arctic is the existence of an unstable situation in the Polar Sea which might rather easily be 'reversed'. Most of the Arctic Ocean is covered by an ice sheet about two to three metres thick, which even in summer does not melt except on its fringes near the continental coasts. The ice is formed from cold water of low salinity which is continually renewed from the large Siberian and Canadian rivers. Underneath there is warmer water of higher salinity – and hence higher density – brought by the Gulf stream. The ice reflects the solar radiation so effectively that even in summer the Artic region is only slightly warmed. However, were the ice to melt, then this top layer would become much warmer and mix with the warmer water coming from the Atlantic. If such a situation arose, then the ice sheet would probably not be renewed even in winter.[49] The idea of transporting some of the water from the Ob and Yenisšei into the arid areas of Central Asia, discussed in the Soviet Union for a long time, might contribute to a diminuation or even a removal of the cold-water 'lid' and hence to the melting of the ice sheet.[50] The expected consequences would be more evaporation, more snow and basic changes in the circulation patterns of the northern hemisphere even if it is impossible to describe these effects in detail. 'We can't even say with confidence whether glaciation would increase or decrease,' said William Kellogg.[51]

Worries of a different type concern the south polar region which is a continent covered by ice some thousands of metres thick. If this huge mass of ice were to melt then the sea level would rise by at least eighty metres. However, this is a process that would take several thousands of years and it is not likely to occur unless there were a warming of the order of twenty degrees. Hence it constitutes no immediate danger. Most of this ice rests on the flat east side of Antarctica. On the mountainous western side of Antarctica, the glaciers rest at least in part on cliffs that are below sea level. Some geologists are of the opinion that a global warming of the ocean waters might result in these waters acting as a sort of 'lubricant' in such places and induce the glaciers to 'slide' (in a period of one to two hundred years) into the sea. This would raise the sea level on a worldwide scale by about six metres and would greatly increase the area of the Antarctic Ocean covered by shelf ice which in turn would affect the Earth's reflection characteristics and result in climatic changes that are hard to predict.[52]

4

Nitrogen and ozone – unexpected relationships

The origin of nitrogen self-supporters

Nitrogen is permanently circulating through living matter. Almost all nitrogen atoms presently found in the atmosphere have repeatedly been incorporated in living organisms and then released. However, even if there had never been life on Earth, our atmosphere would still contain large amounts of free nitrogen. Even if all life were destroyed, causing the atmospheric oxygen to disappear within several million years, due to chemical reactions with the Earth's crust, free nitrogen would remain in the atmosphere.

There is some dispute as to whether the atmosphere already contained appreciable amounts of nitrogen in the beginning of the Earth's history. This depends on how 'reducing' the primeval atmosphere was, i.e., how much hydrogen it contained. At least a fraction of the nitrogen is likely to have been bound to hydrogen forming ammonia (NH_3) which, being readily soluble in water, was an important constituent of the primordial soup (see page 39). The macro-molecules such as protein and nucleic acids that were required for the origin of life, contain nitrogen as well as carbon, hydrogen, and oxygen. Because of the reducing environmental conditions at that time, this nitrogen is not bound to oxygen but mainly to hydrogen and carbon.

As was discussed in part 2, the first living organisms probably originated by chance, when matching building blocks came together. Such matching components, however, were certainly available in the primordial soup in only limited amounts. To be able to propagate, the primeval organisms then had to learn how to build their macro-molecules themselves from more simple components, such as amino acids. Of course, the term 'learn' is meant in a metaphorical sense. Those organisms, enabled by the chance of a favourable mutation (see page 41) to maintain themselves by certain reactions building organic substance, were certainly better off than those that could not achieve this. They could propagate faster and advance to environments where no complete macromolecules existed. Thus step by step organisms evolved capable of producing their nitrogen-containing components from simpler and simpler substances, until finally no complete building blocks like amino acids were required, but merely simple inorganic compounds like ammonia that were still prevalent at that time.

But gradually this substance too became scarce in the primordial soup. On the one hand, the propagation of living organisms caused an increasing demand for nitrogen compounds, while on the other, the transition from a reducing to a neutral atmosphere (see page 30) consequently led to a reduction in the amount of ammonia and an increase in free nitrogen. At first, the primeval organisms that had developed in an at least slightly reducing environment could not utilise free nitrogen. But the highly sought after ammonia became more and more scarce and thus 'more expensive'. Hence the continuance of life was at stake unless at least a few species would be able to learn how to produce these essential compounds from the abundant and thus 'cheap' basic substances, water and nitrogen.

'At that time a situation must have developed when, for certain primeval organisms, no price limit existed to buy ammonia,' says Engelbert Broda, a biophysicist and nuclear chemist at Vienna University, who has been engaged in the research of the nitrogen cycle. He refers to the fact that certain primitive bacteria spend enormous amounts of cash for energy, up to twenty-four molecules of ATP (see part 2, note 33) to convert one single nitrogen molecule into ammonia. This high expenditure of energy is mainly due to the fact that the microorganisms have first to split water molecules in order to obtain the hydrogen required for the production of ammonia.[1] The synthesis of ammonia (NH_3) from nitrogen and hydrogen is a process which in theory liberates energy.

The ammonia factory of microorganisms is a complex system of enzymes that contain iron and molybden, the so-called nitrogenase. Its structure and the way it functions has not yet been eludicated in full detail.[2]

'Certainly the development of such an elaborate and complicated apparatus did not start before it was really necessary, hence before ammonia became scarce,' says Broda. 'But on the other hand, since the system is so complicated, its evolution must have required quite a long time, some hundred million years I would guess. During this time span, various microorganisms had to survive and the system that was evolving to produce ammonia was probably not very efficient or even capable of meeting their own nitrogen requirements. Thus, they had to fall back upon supplementary ammonia, remnants of which were still in the primordial soup. From this one can conclude that the transition from a reducing atmosphere to a neutral one, and hence the corresponding diminuation of ammonia in the atmosphere, must have required a long time.'[3]

Nevertheless, Broda believes that 'nitrogen fixation', i.e. the capability of synthesising ammonia, is likely to be older than photosynthesis. 'There are very primitive bacteria of the genus *Clostridia* that gain their energy exclusively through fermentation,' he explains. 'But while, as far as energy production is concerned, they are capable of performing only relatively simple processes on the side of energy consumption they accomplish the difficult problem of nitrogen fixation evidently because this must have already been a question of life or death at an early stage of evolution.'

At any rate nitrogenase must have developed long before respiration was invented. The complex of the enzymes, being extremely sensitive to oxygen, is likely to have originated long before an oxidising atmosphere existed. Respiring organisms that are also capable of fixing nitrogen possess special built-in safeguards to protect the nitrogenase from oxygen.

When photosynthesis developed, two classes of organisms evolved: those capable of producing their own energetic fuel (i.e. carbohydrates), and those dependent on taking up energy providing food. The evolution of nitrogen fixation led to an analogous classification. On the one side there are self-supporters that produce ammonia from free nitrogen and on the other side there are organisms that can only exist when ammonia or other nitrogen compounds are present in the environment. Many of the latter species meet their nitrogen demands by joining in symbiosis with organisms capable of fixing nitrogen. Others live on nitrogen compounds produced either by metabolic and decay processes or resulting from nonbiological processes such as lightening discharges or ultraviolet radiation.

Common ancestor or natural gene transfer?

Nitrogen fixation and photosynthesis are very old techniques that were already invented when the only organisms that existed were the so-called prokaryotes (see page 44). But whereas the capability of living on photosynthesis was later inherited also by the eukaryotic green plants, all the known organisms able to fix nitrogen are primitive unicellular prokaryotes. One cannot explain why this is the case, but it is hard to believe it is just a matter of chance.

One usually assumes that organisms having in common a certain system of enzymes and hence certain common biochemical properties are closely related. This is obviously not the case with the species able to fix nitrogen. Apart from the fact that they all are prokaryotes they hardly have anything else in common. They belong to different families and classes of bacteria and blue-green algae. Almost all of them have close relatives that are not capable of fixing nitrogen.

One can hardly imagine that a complicated system of enzymes such as the nitrogenase could have evolved independently in several dozen different classes of species. It is all the more unlikely since the enzyme systems of the different organisms living on nitrogen fixation show many similarities, although they are not completely identical. Thus a branching at a later state of evolution appears more likely than independent parallel evolution by chance. But if all organisms living on nitrogen fixation descend from the one common ancestor that invented this technique, then also all their close relatives that are not capable of managing this technique must be descendants of just this same ancestor. How can one explain that so many of these descendants have lost such a useful and advantageous ability?

'I think this might well be the case,' E. Broda replies on this question, 'since nitrogen fixation involves, although being advantageous, great expense. Some bacteria spend almost half their energy budget in producing ammonia, and nitrogenase amounts to nearly five per cent of their cell protein. This is quite a high percentage for an enzyme that only serves as a catalyst.'

Thus in a metaphorical sense, one could say that running one's own ammonia factory requires a high capital investment as well as high operating costs. But once organisms existed that were capable of producing ammonia, it could thus be advantageous for other species to do without such a costly installation, and rather spend their energy resources on other things and somehow manage to acquire nitrogen compounds from those maintaining fixation.

J. R. Postgate, a British biochemist, suggested another possible explanation. He succeeded in transferring the capability of nitrogen fixation from the Klebsiella bacteria, a species found in the human intestine, to Coli bacteria that live in the same environment and that cannot fix nitrogen under natural conditions.[4] This is in principle analogous to the process through which the power of resistance against certain drugs is transferred between different species of pathogenetic agents. Based on his experiments, Postgate suspected that similar processes of gene transfer through packages of nucleic acid (plasmids) or viruses might have occurred in nature. That way, the capability of producing nitrogenase might have been transferred from one ancestor to different species not related to each other.

Broda is sceptical: 'Klebsiella and Coli bacteria are rather close relatives,' he remarks. 'Thus a gene transfer is considerably easier than among organisms that are not so closely related with respect to their evolutionary history. Moreover, nitrogenase is a large and complex system. Its production does not only require just one single gene but a combination of many genes. I don't know of any case in nature where such a large package of genes was transferred among only distant relatives.'

Among the some 200 known genera of bacteria, classified in ten families, at least 25 genera from three families comprise certain species capable of fixing nitrogen. As a rule, only few and not all species of a genus show this ability. Among 15 more genera scientists expect to discover species that fix nitrogen, but research in this field is still in the early stages. Among the 165 existing genera of blue-green algae, species able to fix nitrogen were found in 23 genera. They belong to two of the three families of blue-green algae.[5]

In the deeper layers of the soil, where oxygen is deficient, the most important nitrogen fixers are the aforementioned anaerobic *Clostridia* that thrive all over the Earth. In ventilated topsoil layers, nitrogen fixation is mostly maintained by respiring *Azoto* bacteria, one of the few genera that exclusively consist of nitrogen fixers. In extreme climatic conditions, such as prevail in deserts or on high mountains, and in the polar regions, it is mostly the fixing blue-green algae – either alone or as lichen in symbiosis with fungi – that play a pioneering role for making these areas habitable for vegetation.

Nitrogen fixing bacteria, algae or lichen usually also settle on the surface of multicellular plants, on their leaves and stalks, on the bark or in the root domain. According to Whittaker, rain water falling on a forest contains more dissolved nitrogen at the height of the tree tops than that trickling away from the plants on the ground.[6] The nitrogen is extracted by microorganisms from the water running down the plants, and it is assumed that finally a fraction of the ammonia produced by the nitrogen fixers is somehow taken up by the host plant. Similarly one can assume that nitrogen fixing bacteria populating the digestive tract of humans and animals[7] somehow contribute to the nitrogen (protein) supply of their hosts' organisms.

Besides such comparably loose associations there are also genuine highly organised and close symbioses, where nitrogen fixers supply the desired ammonia to their partner that in exchange provides energetic nutrients or other advantages. Often the partner provides an environment deficient in oxygen that is obviously advantageous for the maximum production of ammonia. Apparently, such an environment is preferred even by those nitrogen fixers that are also capable of respiring.

The best-known of such symbiosis are the aforementioned lichens (blue-green algae and fungi) and that of leguminous plants with rhizobia populating their roots. Where rotational cropping is practised, peas, beans, clover, lucerne, soybeans, peanuts and other leguminous plants are often used because part of the nitrogen compounds produced by the rhizobia enter the soil and thus enrich it.

Alder trees are known to enter in a similar symbiosis with another species of nitrogen fixing bacteria. Additional nitrogen is fed to the soil since, compared to other deciduous trees growing in temperate climates, the alder produces a very large number of leaves with an exceptionally high nitrogen content. For rice farming in India a fern living in symbiosis with a fixing blue-green algae is important. In Brazil symbiosis of certain kinds of maize and sugar-cane with nitrogen fixing bacteria were discovered by J. Döbereiner.[8]

Many of the microorganisms fixing nitrogen, and many kinds of symbiosis between higher plants and nitrogen fixers, have only been discovered during the past few years. Many more such discoveries can be expected. Obviously there are considerably more nitrogen fixers and much more nitrogen is fixed, than was generally assumed a few years ago. Nevertheless, nitrogen compounds usable for higher plants are a scarce commodity in many environments.

For agriculture it would be highly desirable to develop strains of grain capable of being self-supporting with regard to nitrogen. There are research groups trying to breed new species of nitrogen bacteria that can enter into a symbiosis with grain or new sorts of grain that can live in symbiosis with rhizobia. There are also experiments for directly transferring nitrogenase genes to grain.[9] A breakthrough along any of these lines might be an important contribution to improving the global food situation.

The nitrogen cycle

Protein, nucleic acids and other vitally important substances are produced from less complex organic (carbon-containing) nitrogen compounds. In most environments these compounds are scarce. Thus, whenever they become available through metabolic or decay processes, they are eagerly taken up and incorporated by many organisms. Therefore since very early times there has been a cycle with nitrogen compounds being immediately utilised by other organisms rather than being decomposed into basic substances.

For the synthesis of macro-molecules, however, energy is required. At the beginning lightening discharges and ultraviolet radiation provided the energy necessary for the formation of the substances dissolved in the primordial soup. Later on the organisms themselves produced the energy required for producing higher nitrogen compounds from single basic substances. When such compounds decay, this energy is released. Thus, nitrogen compounds can serve not only as building materials but also as fuel, and wherever nature provides the chance of acquiring energy, generally organisms develop to make use of this possibility.

Most likely already in the days of yore, even before nitrogen fixers existed, microorganisms must have evolved that gained their energy from the decomposition of organic nitrogen compounds from other organisms. As long as the atmosphere was still reducing, the end product of such cycles could only be ammonia (and not free nitrogen). Despite changed environmental conditions such ammonia producing bacteria have survived to the present day, and the ammonia they release serves as a building material for other organisms.

Using organic nitrogen compounds as fuel is not only the privilege of bacteria that are specialised in this. We human beings for instance can decompose excess protein that was taken up with the food we eat and which is not required as a building material. Our organism gains energy through this process, though not in a very efficient way.

With the transition to a neutral atmosphere and the evolution of nitrogen fixers, free nitrogen was also incorporated in the biological cycle. However, throughout the age of blue-green algae, for almost two thousand million years, evidently no biological processes existed through which organic nitrogen compounds could be decomposed with free nitrogen being the end product. No microorganisms are known that convert ammonia to nitrogen. Throughout this long time span, the cycle could probably only be closed with ammonia being released into the atmosphere and there being decomposed by ultraviolet radiation. Thus during the age of blue-green algae no real equilibrium state is likely to have existed for either nitrogen or carbon. Apparently, more nitrogen was incorportated into the biosphere by fixing organisms than released from escaping ammonia. Thus, the biomass increased, and considerable amounts of organic nitrogen compounds are likely to have entered the sediments.

Only when oxygen began to accumulate in the oceans and in the atmosphere was there the possibility to use ammonia itself, the previous end product of the decomposition of organic nitrogen compounds, as fuel in a chain of oxidation reactions. Nitrate (NO_3^-), the highest oxidised nitrogen compound is the end product of this metabolic chain of so-called nitrifying microorganisms (*Nitrosomonas*, *Nitrobacter*). Since our atmosphere contains oxygen, oxides of nitrogen are also formed through inorganic processes such as lightening. Through precipitation, these are transferred to the soil and water and thus constitute a natural nitrogen fertilisation.

Under the environmental conditions as existing on Earth today, free nitrogen and free oxygen are not stable together. In principle, as was pointed out in the introduction, they are likely to react with each other until all the oxygen is used up. These reactions occur in fact when triggered by a certain amount of electrical or thermal energy. Nevertheless, nitrates cannot accumulate in the soil or in the waters, because, as soon as oxygen had appeared in the atmosphere, most of the plants that require nitrogen compounds but cannot cope with free nitrogen had learnt to utilise nitrates. From these they produce ammonia, the basic substance for building higher nitrogen compounds, with carbon dioxide being formed as a by-product.

Only in areas where plants do not grow, for instance in the seabed or in deserts, could nitrates accumulate. However there are also microorganisms that decompose nitrates (see page 59) by nitrate respiration (so-called denitrifying bacteria). As long as sufficient oxygen is available, they use organic carbon compounds as fuel and gain their energy through respiration. In an environment rich in nutrients but deficient in oxygen they can get oxygen from nitrates, in order to use it for the oxidation of carbon. These microorganisms recycle the oxygen that is bound in nitrates back into the atmosphere as carbon dioxide. Thus they generally prevent the accumulation of nitrates even in areas where no plants exist.

Since nitrate respiration is more expensive than ordinary res-

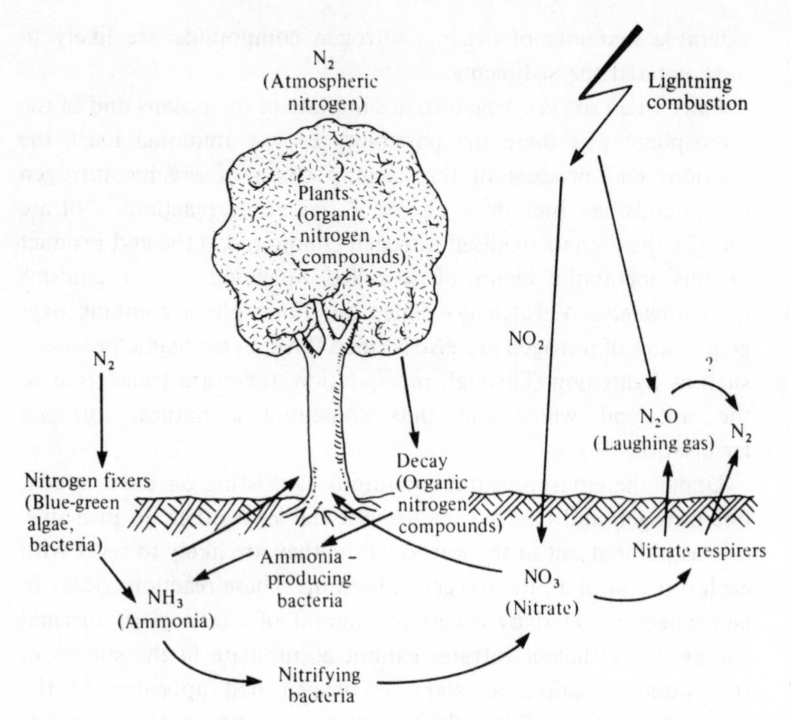

Figure 15 Simplified scheme of the nitrogen cycle. A fraction of the laughing gas (N_2O) is decomposed in the stratosphere. Another fraction is assumed to be converted into nitrogen in the near-surface layers or by microorganisms through a mechanism that is as yet unknown. (See p. 121).

piration, denitrifying bacteria do not use more oxygen from nitrates than is really necessary. As long as the atmospheric oxygen content is more than five per cent they live on ordinary respiration. Below this level, they switch over to nitrate respiration. At this stage only part of the oxygen is extracted from the nitrate with laughing gas (N_2O) being the end product.[10] Only in a situation of extreme oxygen starvation is all the oxygen that is required by these organisms taken from the nitrate, with the end product in this case being nitrogen.

With the origin of nitrate respiration, the biological cycle became closed. At the end of a long chain of reactions, organic nitrogen compounds could now be decomposed completely, and free nitrogen, the most simple basic substance, could be recycled to the atmosphere. Generally, it is assumed that this cycle is balanced under natural conditions, undisturbed by human activities, but this has not yet been proved. Many details of the nitrogen cycle are as yet insufficiently known, and the figures published by various scientists still show considerable scatter, as is illustrated by the following table.

Flow rates of the nitrogen cycle.

(Comparison of figures published by different authors, in million tonnes per year.)

	I	II	III
Biological fixation	129	265	180
Industrial fixation	40	40	40
Nitrogen compounds in precipitation*	79	200	240
Total denitrification	160	315	349
N_2O from denitrification	?	70	60

I C. C. Delwiche, G. E. Likens in: W. Stumm (Ed.), op. cit., p. 79.
II J. Hahn, C. E. Junge, op. cit.
III M. B. McElroy *et al.*, *Rev. Geophys. Space Phys.* **14**, 143, 1976.
* Produced by lightning, combustion and other processes.

Intentional and unintentional nitrogen fertilisation

In a natural ecosystem, a primeval forest for instance, the nitrogen budget is normally balanced. The total mass of nitrogen compounds carried off to the rivers through surface and ground waters is small. Generally, the losses due to this washing out and due to bacterial nitrate respiration can be compensated for by nitrogen fixing organisms and compounds dissolved in rain water.

If a forest is cleared the mass of nitrogen compounds and other vitally important substances washed out to the rivers increases tremendously.[11] This by itself gradually causes impoverishment of the soil, but agricultural utilisation makes even higher demands. When harvested, the nitrogen compounds contained in the crops are removed from the local ecosystem. If the agricultural products are used as fodder, at least part of these compounds are fed back to the fields in the form of natural manure. In general, however, most of the crops are brought to the markets in the cities or even exported and the waste products of human consumption are usually not carried to the soil, but rather to a sewage system, the rivers and finally to the ocean. As far as hygiene is concerned, this kind of waste product removal undoubtedly has its advantages. In regions with efficient sewage systems infections with tapeworms, trichinas, hookworms and other parasites and pathogens caught from faeces have nearly disappeared. For the soil however, this means permanent deprivation of nutrients that have to be replaced somehow.

For thousands of years man has, without being fully aware of the mechanism, used farming techniques to counteract the deprivation of the soil. Besides the natural manure, fallowing to allow the soil to 'recover', and the rotation of crops with leguminous plants are some of the methods used. The recovery essentially reflects the accumulation of nitrogen compounds produced by microorganisms that can propagate in the soil during the fallowing period. Likewise the growth of leguminous plants that live in symbiosis with nitrogen fixing bacteria yield, even more effectively, the same result. It is only with the using of natural manure as a fertiliser that almost *all* nutrients are recycled back into the soil. Fallowing and the rotation of crops with leguminous plants can only increase the nitrogen content of the soil, while the other essential nutrients (such as phosphates or potassium) are not replenished. Thus good yields of leguminous plants can often be achieved only if the other missing substances are supplied.[12]

Since it is fairly well known what kind of substances farm products take away from the soil, one can supply those as industrially produced fertilisers and thus get considerably better yields than even on good natural soil. Without the massive application of fertilisers the present world food situation would be even worse. The rejection of fertilisers has somehow become fashionable among supporters of the so-called 'natural farming'. But as a matter of fact *every* kind of agriculture is 'unnatural' and constitutes a drastic interference with the natural ecological system. Whether artificially created farmland is supplied with nitrogen from the growing of leguminous plants or by applying industrially produced fertilisers are mere shades in the degree of human interference.

For most nutrients the flow of the natural geological cycle is accelerated and channelled when fertilisers are applied. More phosphates or potassium salts are added to the soil than would be supplied through natural processes such as weathering, erosion and alluvial deposit. It is only with nitrogen that man imitates industrially a biological process. Ammonia is produced from atmospheric nitrogen (Haber-Bosch process), in the same way as microorganisms fix nitrogen. Depending on local conditions, nitrogen fertilisers are added to the soil either as salts of ammonia or as nitrates, both having advantages and disadvantages. Ammonia salts are adsorbed relatively easily by the constituents of the soil and are thus hard to get for the plants. For nitrates this difficulty does not exist. However, being soluble in water these substances are washed away easily. Moreover, they have to be converted to the required ammonia by the plant itself, involving the consumption of carbohydrates and hence at the expense of part of the harvest yield.

Besides this intentional, there is also 'unintentional' fertilisation by oxides of nitrogen, produced through combustion processes. 'Only a small fraction originates from the oxidation of the nitrogen contained in the fuel,' says Jürgen Hahn who deals with problems related to the nitrogen cycle at the Max-Planck-Institut für Chemie at Mainz. 'The oxides of nitrogen are predominantly produced by the oxidation of atmospheric nitrogen that has entered the engine of a car, the heating system of a power station, of a manufacturing plant or of a dwelling, along with the oxygen required for combustion.' Through combustion, nitrogen dioxide (NO_2) and nitric oxide (NO) can be produced and these form rather strong acids with rain water thus increasing the corroding effect of precipitation. Acidic rainfall, the terror of many industrialised countries, is partly caused by such oxides of nitrogen, though most of the acid in rain water originates from sulphur dioxide that is released into the atmosphere.[13] For man and animals alike the acids formed from the oxides of nitrogen are poisonous. For the soil, however, they constitute a considerable additional amount of nitrogen fertilisation.

Besides the aforementioned oxides of nitrogen, less oxidized laughing gas (N_2O) is also produced in the combustion process. In the near-surface layers of the atmosphere this compound is nearly inert – 'almost like a noble gas,' says J. Hahn. If it enters the stratosphere, however, it contributes to the decomposition of ozone. Additionally, laughing gas contributes to the greenhouse effect as mentioned before (see page 100) and hence to the warming of the Earth. An increase in the abundance of atmospheric laughing gas might therefore constitute a serious risk to the environment, as will be discussed later.

The higher the temperature in the combustion process, the more oxides of nitrogen are formed. On the other hand, higher combustion temperatures imply a higher efficiency of the combustion plant and thus a reduced burden on the environment through carbon dioxide and other combustion products, for the same power output. Thus with regard to protecting the environment two opposing effects result, of which only one can be reduced but at the expense of the other. Usually for reasons of economy, higher temperatures are preferred and this means putting up with heavier burdens from the oxides of nitrogen.

As to the relative proportions of natural and man-made nitrogen fixation no exact figures are as yet available. In 1975 about forty million tonnes of nitrogen were fixed through fertiliser production, and it is estimated that annual rate of fixation will exceed by far a hundred million tonnes at the end of this century.[14] At present some twenty to forty million tonnes of nitrogen are likely to be oxidised every year through combustion processes, and these figures are expected to grow in the future too. Moreover, considerable amounts of nitrogen are added to the biosphere by increased cultivation of leguminous plants. At the Dahlem conference C. Delwiche told me that considering all of these processes together, more nitrogen is already being added to the soil by man than is added through the natural cycles. Other scientists think that Delwiche underestimates the rate of natural nitrogen fixation; according to their estimates something like two to three hundred million tonnes of nitrogen are fixed by microorganisms every year.[15] (See also the previous table on page 115.)

But even according to the most conservative estimates, man is at present increasing the amount that enters the soil by some twenty to thirty per cent, and this percentage is likely to at least double by the turn of the century. This must be having a massive impact upon the natural cycle, and we cannot take it for granted that the denitrifying bacteria can keep pace with this development. If they cannot then we have to expect an increase of the nitrate concentration in subsoil waters, as has already been observed in different parts of the world, and poisonous concentrations of these compounds may occur in drinking water.[16] If they can, however, increased nitrate respiration will increase the concentration of laughing gas in the atmosphere with possible consequences no less critical for our environment.

Almost a noble gas

'Nitrous oxide (N_2O)', says the *Encyclopaedia Britannica*, 'is a colourless gas with a pleasant sweetish odour and taste; ... it has anaesthetic properties in some cases preceded by peculiar effects that led to its name laughing gas.'

According to the most recent measurements, the air we breathe contains about 30 millionths of a per cent (by volume) of this gas, which is equivalent to a total mass of 1.5 thousand million tonnes.[17] At present, the nitrous oxide content of the atmosphere is likely to be already higher than it was at the beginning of this century. This cannot be proved however, since instruments capable of measuring such low concentrations with sufficient accuracy did not exist before World War II, and even the first measurements dating back to the 1950s have to be considered as unreliable.

In granaries and greenhouses distinctly higher levels of nitrous oxide are measured, but nowhere have concentrations ever reached levels that might have noticeable effects on human beings. If mankind were to experience a tenfold or even hundredfold increase of the nitrous oxide concentration in air, nobody would be seized with laughter. But even a mere doubling or trebling of the present concentration might cause a situation that will not be at all hilarious.

'Nitrous oxide is fed to the atmosphere from the continents as well as from the oceans,' says J. Hahn. 'The annual production rate is not known exactly. Estimates range between fifty and five hundred million tonnes, with the lower values being preferred at present. The most important source is the production by micro-organisms through nitrate respiration. We suppose however, that in other biological processes, in particular through the activity of nitrifying bacteria, nitrous oxide is generated as a by-product too, but this has not been proved so far.'

Based on laboratory experiments, the American E. C. Zipf estimates the total amount of laughing gas produced by lightning to amount to fifteen to ninety million tonnes per year.[18] Several million tonnes of nitrous oxide are also formed by combustion processes, and the chemical industry adds smaller amounts to the environment too. 'As far as I know, the production of laughing gas by sewage disposal works has not yet been investigated,' Hahn commented. 'However I can imagine it to be considerable.'

There are several factors determining the production rate of nitrous oxide in the soil: the ventilation depending in turn on the humidity of the soil; its chemical composition; and its fertility. But even extremely barren desert lands exude small quantities of nitrous oxide. The highest production rates of laughing gas are found in farmland after nitrogen fertiliser has been added. As a rule some 20 to 30 per cent of the fertiliser is decomposed by nitrate respiration before the plants can take it up. Among the products of this decomposition, the fraction of laughing gas varies between twenty and sixty per cent. On average it is about one third, while the remaining two thirds are released as nitrogen.

Nitrogen compounds taken up by the plants are also decomposed in the course of the biological cycle, and thus get into the effective range of nitrate respiration. Similarly, not only the fertilisers are decomposed by the various nitrogen bacteria, but also manure and the nitrogen compounds produced by rhizobia in symbiosis with leguminous plants. 'One can say that basically *each* increase in the total mass of nitrogen circulating in the biosphere will sooner or later result in an increased rate of nitrous oxide release into the atmosphere,' explains Hahn.

'And what is actually happening with the laughing gas in the atmosphere?'

'Well, that is something we too would like to know. At present there is no really satisfying answer to this question. As I told you before, nitrous oxide behaves almost like a noble gas in the troposphere, i.e. in the lowest layer of the atmosphere. No chemical reactions binding or destroying it are known. If it enters the stratosphere it is decomposed and the product of this decomposition (NO) then destroys ozone. But the amount that is transferred to the stratosphere every year seems to be *considerably* smaller than the estimated annual production. According to Junge, it amounts to twenty-eight million tonnes per year, and independent assessments by some American groups even estimate it to be a mere fourteen million tonnes,[19] while the annual production, as mentioned before, is likely to be more, perhaps much more than fifty million tonnes. Thus there is probably another as yet unknown sink mechanism for nitrous oxide, a process or perhaps a series of processes through which this gas is decomposed again.'

This unknown sink is unlikely to be found in the oceans, Hahn continues, since all investigations carried out so far have shown that in different oceans and at various geographic latitudes, seawater is supersaturated with laughing gas as compared to the air above. Per unit area, oceans certainly produce, on the average, distinctly less nitrous oxide than the soil, but they definitely emit certain amounts of this gas into the atmosphere, rather than absorbing it from the air. In the troposphere, a very slow decomposition by light might perhaps be possible, but only small quantities of nitrous oxide could be destroyed that way, and certainly much less than in the stratosphere. Thus, the search for the unknown sink of nitrous oxide, i.e. processes that destroy this gas in the near-surface layers or in the soil, might lead to a species of yet unknown microorganisms specialising in the decomposition of laughing gas. Delwiche thinks that in marshy soils deficient in oxygen, possibly in the tundra, certain bacteria might utilise atmospheric nitrous oxide as a source of oxygen.[20] But for the time being this remains a mere speculation.

'The elucidation of this problem, as repeatedly pointed out by our Institute, is of utmost importance,' says Hahn. 'Otherwise, it may well be possible that man, without knowing about it, may damage this sink, and thereby the future burden of the atmosphere might be even further increased.'

It is very difficult to assess the future trend of the atmospheric nitrous oxide concentration as long as the mechanism of this sink is not known. One also needs to know whether this sink will grow with increasing nitrous oxide production – this would be likely in the case of a nonbiological decomposition process – or whether it will remain constant – likely to be the case if decomposition is achieved by microorganisms and the propagation of these species if limited by other circumstances. A further problem exists in not knowing the trend in atmospheric nitrous oxide concentration in the past. As opposed to carbon dioxide one cannot test the reliability of the models by just feeding them with data from past decades in order to verify whether they can 'predict' present levels reasonably well. Moreover basic data like trend predictions for the future usage of fertilisers show considerable scatter.

Despite all these uncertainties one simply cannot do without a prognosis, since the consequences of an excessive concentration of laughing gas in the atmosphere might be no less critical than those of increasing carbon dioxide. Again one has to solve an equation with many unknowns – and try to assess what would happen if . . .

Such an attempt was made at the Mainz Institute.[21] In this the optimistic assumption was made that decomposition (the sinks) and production (the sources) of laughing gas will grow at the same rate. For the usage of fertilisers an annual growth rate of 6 per cent until the year 2000 was assumed, levelling off then at that consumption rate. For the consumption of fossil fuels – since oxides of nitrogen are a by-product of combustion – an annual growth rate of 3.7 per cent until the year 2050 was assumed, with no further increase thereafter.

Perhaps the most important result of these computations was, most surprising to the nonprofessional, that the nitrous oxide content of the atmosphere will increase even if the unknown sink grows at a rate that keeps pace with the source. 'The amount of laughing gas in the atmosphere continues growing until a new equilibrium is reached at a higher level. This occurs some time after the source has levelled off at a constant rate,' explained Hahn, when he showed me these results, at that time unpublished. 'This is because the sink grows, supposing it can grow at all, proportional

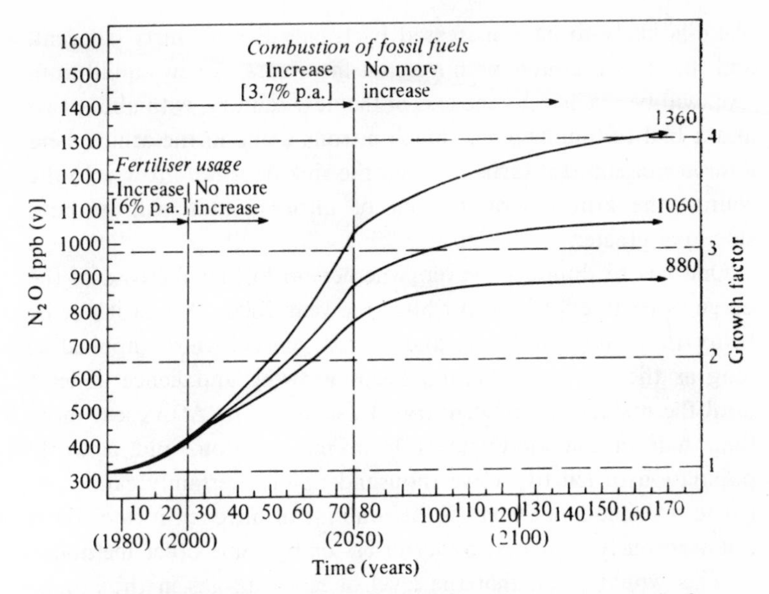

Figure 16 Probable range of increase of atmospheric laughing gas (N_2O) concentration making various sets of reasonable assumptions about increase in fertiliser usage, percentage of fertiliser converted to N_2O, atmospheric residence time, etc. (From J. Hahn in: W. Bach (ed.), *Man's Impact on Climate*, Elsevier (in print).)

to the content of the atmosphere, rather than at a rate proportional to the source.'

'Imagine a tank with a small hole in the bottom,' he continued. 'The more water is in the tank the greater the water pressure and hence the faster the water runs out through the hole. If the tap is almost closed the tank will fill up to a level where inflow and outflow balance. If the tap is opened a little more, the water level in the tank rises until a new equilibrium is established. In fact, if the content of laughing gas in the atmosphere were to remain unchanged the efficiency of the sink would have to grow considerably *faster* than the source. Perhaps this may be possible to achieve by planned human intervention once the mechanism of these decomposition processes is known.'

Based upon these assumptions, the Mainz group predicted that by the year 2000 the concentration of laughing gas in the atmosphere is likely to have increased by twenty-five to thirty per cent, and that it will double within about fifty years. A new equilibrium level will be reached by the end of the next century, with about two and a half to four times as much nitrous oxide in the atmosphere than at present. If it turns out that the sink does not grow with the source, the atmospheric burden of nitrous oxide will be considerably greater.

One has to doubt, however, whether no further increase of the usage of fertilisers will occur after the year 2000. 'Would it not be better to assume that more and more fertilisers will be applied as long as the world population keeps growing and hence at least until the middle of next century?' I asked Hahn. 'At present more than half of mankind suffers from malnutrition, and a world population of ten to twelve thousand million certainly cannot be fed to an adequate extent, if the amount of nitrogen in the soil is not vigorously increased by fertilisers or by some other methods.'

'This would mean that the level of laughing gas in the atmosphere would keep growing even longer and more rapidly than predicted by our model,' he replied, 'and if the worldwide usage of nitrogen fertilizers does not increase by merely six per cent, but perhaps by ten per cent as is now often predicted, nitrous oxide will accumulate even faster still. On the other hand it is possible that in the long-term the consumption of fossil fuels might increase by less than 3.7 per cent and in the second half of the next century it may even decrease rather than level off. This would somewhat reduce the figures for the predicted increase of laughing gas in the atmosphere. But supposing that our estimates of the residence time of nitrous oxide are realistic and that no feedback mechanism exists slowing down its production with increasing abundance, then our model predicts, even for the most optimistic assumptions, that at least doubling of the atmospheric concentration of laughing gas must be expected for the first half of the next century.'

I referred to the fact that agrarian chemists have criticised these forecasts as being 'unrealistic and exaggerated.'[22] Hahn replied to this criticism in this way: 'I can indeed understand that agrarian chemists and also the fertiliser industry in general try to defend themselves to the last, although strictly speaking, they were not attacked at all. The only thing *I* reproach them with is the fact that farmers are encouraged to consume fertilisers at a sometimes unreasonable rate. With vegetable farmers in Middle Europe adding to their fields up to six hundred kilograms of nitrogen per hectare, nobody ought to be surprised any more when waterworks providing drinking water cannot cope with increasing amounts of nitrates in subsoil waters. And nitrous oxide production in the soil is likely to be highly stimulated as well. One cannot simply resolve these problems by just ignoring their existence.'

Spray cans, fertiliser, supersonic aircraft

In the atmosphere, ozone is only a trace gas. Its total mass is less than 3.3 thousand million tonnes. Even in the stratosphere, where the highest concentrations of this triatomic oxygen molecule are observed, its fraction is only some five thousandths of a per cent of the total volume. For this reason alone, small amounts of certain substances can cause a considerable ozone depletion. Spray cans, fertiliser and supersonic aircraft have in common that they all contribute pollutants capable of destroying ozone.

Under natural conditions there is, as was discussed in part 2 of this book, an equilibrium between the formation and the destruction of ozone (see page 36). Oxides of nitrogen, free oxygen atoms, products of photolytic decomposition of water vapour, and some other minor constituents, are involved in the destruction mechanisms. Their relative contributions with respect to ozone decomposition are, according to recent figures prepared by a panel of the American National Academy of Sciences, about sixty per cent, twenty per cent, ten per cent and ten per cent, respectively.[23] Laughing gas is the only one of the oxides of nitrogen produced in the lower regions of the atmosphere that is mixed into the stratosphere. Both nitric oxide and nitrogen dioxide react, as was mentioned before, with water vapour forming acids that are washed out by precipitation. Laughing gas, behaving like a noble gas in the troposphere, reacts with ultraviolet-produced 'excited' oxygen atoms in the stratosphere. Thereby nitric oxide (NO) is formed which reacts in the following chain with ozone:

$$NO + O_3 \rightarrow NO_2 + O_2$$

$$NO_2 + O \rightarrow NO + O_2$$

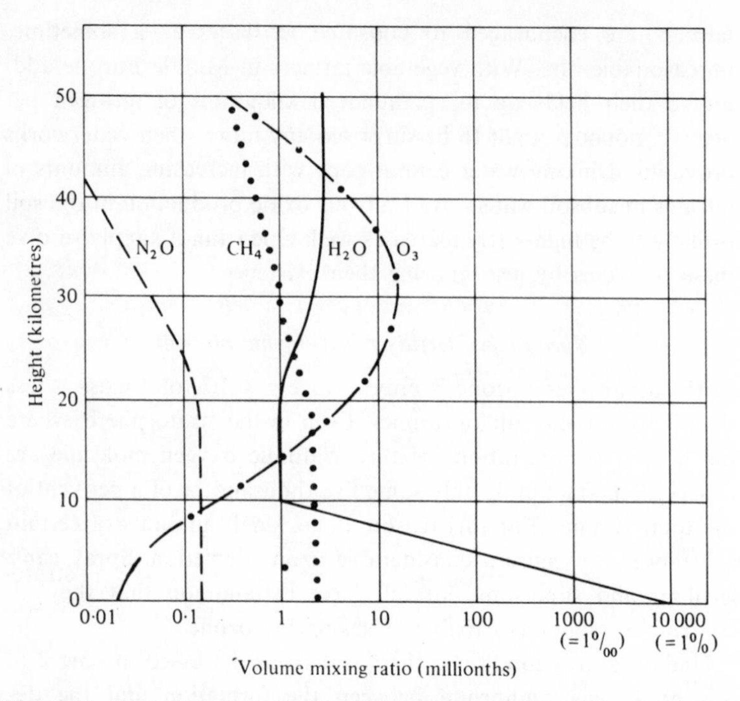

Figure 17 Vertical profiles of atmospheric trace gases. Laughing gas (N_2O) is decomposed in the stratosphere by the process discussed in the text as well as by ultraviolet photolysis. Methane (CH_4) undergoes similar, although weaker, decomposition reactions in the stratosphere. Ozone (O_3) is formed through the reactions discussed on page 36. Water vapour (H_2O) decreases with increasing height in the troposphere due to condensation and ice crystal formation. In mid-latitudes, the upper limit for cloud formation and precipitation processes is about ten to twelve kilometres and in the tropics it is at about eighteen kilometres altitude. Above this 'cold trap', in the stratosphere, only traces of water vapour exist. Some water vapour is produced in the stratosphere in the course of methane decomposition, as indicated by the slight increase above twenty kilometres. (The distribution of water vapour is not well known as yet, and the available data show considerable scatter.)

As a net result of both reactions, one ozone molecule (O_3) and one free oxygen atom (O) are converted into two ordinary oxygen molecules (O_2), while nitric oxide (NO) is restored at the end of the second reaction ready to enter the cycle again and again. Thus in this reaction chain, nitric oxide acts as a catalyst, i.e. one nitric oxide molecule can decompose many ozone molecules, until it is converted into a chemical compound that does not react with ozone – as for instance through a reaction with products of photolytic decomposition of water vapour.

If the abundance of laughing gas in the atmosphere increases due to the increased usage of nitrogen fertilisers, larger amounts of this gas will get into the stratosphere. This is likely to cause an increased ozone depletion. It is very difficult, however, to quantitatively assess these effects, since the different substances that decompose ozone also react with each other.[24] The systematic investigation of the complex scheme of chemical reactions going on in the stratosphere has only just begun in recent years. New interrelations are still being discovered, and the research activities in this field may even yield some surprises. Only a few years ago it was assumed that doubling the nitrous oxide concentration in the atmosphere would cause an ozone depletion of about twenty per cent. This effect is now estimated to cause 'only' two to six per cent ozone depletion.[25]

As mentioned before the increasing usage of nitrogen fertilisers is not the only human activity likely to cause a threat to the ozone layer. Supersonic aircraft dump oxides of nitrogen and water vapour that enter the chemical reaction chains, directly into the stratosphere. As long as not more than some twenty to thirty Concordes or Tupolev-144s are operating in the stratosphere, and the larger Boeing-2707 remains in cold storage, the environmental impact of these aircraft is likely to be negligible. What would happen if several hundred supersonic aircraft were to operate at stratospheric levels is another question, and respective model predictions still differ considerably.[26]

The third threat to the ozone layer, presently recognised as the most dangerous one, is constituted by certain substances called 'halocarbons' used as propellants in spray cans. (Concern is about two species of halocarbons, namely trichlorofluoromethane ($CFCl_3$) and dichlorodifluoromethane (CF_2Cl_2) or, according to the nomenclature of one major producer, Freon-11 and Freon-12, respectively.) During the 1930s these substances were originally developed as refrigerants for cold-storage plants. Substances were needed that boiled between 0 and -40 degrees centigrade, that were chemically stable, non-inflammable, non-poisonous, not too expensive and easy to produce and to store. Thanks to a lucky chance, a group of substances was found that met all these requirements.[27]

At that time, these halocarbons were produced for refrigerators and for fire extinguishers, but because of their properties they were eminently suitable as propellants for spray cans. The first such product was a hair spray put on the market in 1950. Since then the production of spray cans has been increasing rapidly. At present they already consume half the world's production of these halocarbons, which is estimated to amount to about one million tonnes per year.

It is precisely because Freon-11 and Freon-12 are chemically stable in the troposphere that they behave like noble gases. They are not decomposed by chemical or biological processes and thus gradually accumulate in the atmosphere. This in itself would not be a worry but there is concern about two other effects. These compounds increase, as mentioned before, the greenhouse effect and hence the warming of the atmosphere in addition to that caused by the increase of carbon dioxide (see page 100) and they are decomposed in the stratosphere by solar ultraviolet radiation, whereby chlorine atoms are liberated. Through reaction chains analogous to those we have demonstrated for nitric oxide, chlorine reacts with ozone and free oxygen atoms, without being consumed itself. Hence, like a catalyst, it can also induce decomposition of ozone, until it is finally bound to a nonactive compound through a reaction with another species.

Attention was first drawn to this environmental problem by two American scientists, M. J. Molina and F. S. Rowland, in a paper published in 1974.[28] One can understand that the reaction of the chemical industry with regard to this issue was not at all enthusiastic. Slogans such as 'speculation', 'premature conclusions', 'overstatements', 'insufficient experimental data' were raised then, and also objections against intentions to ban spray cans as long as there was not conclusive evidence of their environmental risk.

Meanwhile the panel appointed by the American National Academy of Sciences mentioned previously had thoroughly assessed this problem and concluded that the ozone layer is likely to be reduced by 6 to 7.5 per cent, if the worldwide usage of spray cans continues at the level reached in 1973.[29] According to most recent investigations about coupling reactions in stratospheric chemistry, an even greater ozone depletion of about ten per cent must be expected.[30] For such a constant consumption rate of halocarbons this new equilibrium of the ozone concentration would be established within forty to fifty years, since a considerable time is required for these pollutants to be mixed from the near-surface layers into the stratosphere. For the same reason, even if the usage of halocarbons were to be stopped completely, the ozone depletion would nevertheless continue to increase for at least ten more years, and it would require some sixty-five further years for the depletion to be reduced to half its peak value.[31]

A meeting of the World Meteorological Organisation (WMO) in June 1978 and a workshop sponsored by the United Nations Environment Programme (UNEP) meeting in Bonn in November 1978 were even more pessimistic. Both expected an ozone depletion reaching eventually fifteen per cent if halocarbon releases were to continue indefinitely at the 1977 rate.[32]

When I asked participants of the Dahlem conference, they all agreed that use of halocarbons for spray cans ought to be banned. 'Spray cans are indeed useful, but they are not as vitally important as fertilisers for instance,' Junge replied. 'The arguments against this usage of halocarbons are well founded and convincing. Even if it turns out that the predicted figures of future ozone depletions are too high, which I consider unlikely, it appears to me still much better to take such precautions, rather than taking the risk of endangering the ozone layer.'

The answers given by Bolin and Flohn were along the same lines.

Meanwhile the USA severely restricted the use of halocarbons in spray cans and Sweden has banned it altogether. The EEC countries, especially Britain and France, are however still hesitating to follow suit.[33]

There is basically no objection to the usage of halocarbons in refrigeration or in air conditioners. One would need to ensure however, at best by law, that the substances contained in worn-out equipment are collected so that they cannot be released into the environment.

A slight depletion in the atmospheric ozone layer would, above all, result in increasing the intensity of ultraviolet radiation of wavelengths between 290 and 320 nanometres (millionths of a millimetre) that penetrates to the Earth's surface. In limited intensities this kind of radiation is part of our present-day radiation climate, and it is more intense in high mountainous regions than in the lowlands. In small doses, this radiation causes sunburn, and an exposure for years or decades can cause skin cancer. An even more massive ozone depletion would have far more serious consequences. Then, radiation of wavelengths less than 290 nanometres, which at present are completely absorbed by the ozone layer, would also reach our environment.[34] This radiation is capable of decomposing nucleic acids and protein.

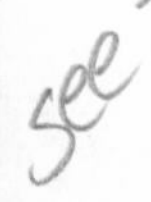

According to a report of the American National Academy of Sciences every one per cent of ozone depletion is likely to correspond to an increase of the occurrence of skin cancer by about two per cent. Thus for an ozone reduction of five per cent, forty thousand additional cases of skin cancer per year must be expected for the USA alone.[35] In most cases however, skin cancer is amenable to successful therapy. Only one particular kind, so-called melanoma (rank growth of dark pigmentary cells), is usually fatal. At present about five thousand deaths due to skin cancer are reported in the USA every year. Thus five per cent ozone depletion and hence ten per cent increase of skin cancer occurrence is likely to cause five hundred additional fatal disease cases every year in the USA alone.

One cannot extrapolate such estimates for the total world population by simple multiplication since white people are not only more susceptible to sunburn but also more susceptible to skin cancer than are coloured people. There is no doubt however that a noticeable and statistically verifiable consequence must be expected everywhere on Earth.

In view of these problems, one can certainly do without spray cans and also, if necessary, without supersonic aircraft (although the industries concerned will undoubtedly resist). But if we want to ensure the food for the rapidly growing world population we certainly cannot do without fertilisers and/or other methods of increasing fixed nitrogen in the soil – and hence a higher abundance of laughing gas in the atmosphere. Thus will there be another alternative for next centuries apart from choosing between starvation or skin cancer?

By posing such a question one should not lose sight of the relative orders of magnitude. As mentioned before skin cancer can be cured in most cases. Even if the number of fatal cases were to double this would still remain a small fraction of the number of people that starve to death or die because of malnutrition. *Already today* hunger is undoubtedly a much more dreadful scourge for the human race than the future consequences of the inevitable increase in fertiliser usage.

The rule of thumb, that an ozone depletion by one per cent is likely to cause a two per cent increase of skin cancer occurrence, will apply only if *no precautions are taken.* It is relatively easy though to protect oneself against the impact of ultraviolet radiation. If the radiation dose were only slightly increased, it would be probably sufficient to oil or cream the body continuously, to wear a straw-hat on sunny days and not to expose the bare skin to the Sun. This is essentially the same precaution that is taken by the experienced in high mountainous regions, where ultraviolet radiation is more intense. It is not clear, however, whether such precautions would be sufficient in case of a severe enhancement in the radiation dose.

Probably the most severe *immediate* risk for man is his tendency to underestimate such dangers. Yet ozone depletion and the associated enhancement of ultraviolet radiation dose will proceed slowly, and skin cancer due to the radiation burden often shows up only after decades. Will people really remember always to oil themselves and their children several times every day? Will people really do without intensive sun-baths – which in the view of many doctors are not at all harmless even with present-day radiation conditions? There is certainly much education work that remains to be done.

Of course enhanced ultraviolet radiation is not only dangerous for man but also for other organisms. In the scientific literature dealing with the ozone problem, the question is often raised, what would be the impact of enhanced radiation on crops, on insects important for pollination or on plankton and microbes? But nobody seems to know the answer.[36]

The results of relevant experiments carried out so far are controversial, reported Harvey Brooks to the Dahlem conference. This American physicist, who is particularly involved in studying the relationships of science and society, thinks that some more decades might have to elapse before fairly reliable statements on such effects can be made.[37]

'I think one should not overrate these potential threats,' commented E. Broda. 'If the number of fatal diseases due to skin cancer were to double, this implies that, on a global scale, some ten thousand more people would die every year, and this is a serious problem indeed. If the rate of fatal diseases of cattle were to double however, from one to two in a hundred thousand, say – and such orders of magnitude are likely to be realistic – even cattle farmers would not really have to worry very much about losses. If it were to turn out that there are no important species of animals or plants, that are at least a hundred times as susceptible to the impact of ultraviolet radiation than the human race, *this* problem can certainly be considered as unimportant.'

'And what about algae, bacteria and other microorganisms?' I asked. 'There is no doubt that these are considerably more susceptible to radiation than multicellular organisms.'

'Certainly, but only if they are directly exposed to the radiation. Soil bacteria from instance are well protected. In water, only the plankton populating the uppermost few centimetres would be concerned. Microbes in the atmosphere or living on the surface of plants are in greater danger. Many are likely to perish, but I cannot imagine these species dying out that way. There would always be a sufficient number of individuals surviving, and microbes propagate at an extremely fast rate. A few days of bad weather and hence less radiation would be sufficient to restore the population.'

Nevertheless, Broda remarked, one should not push aside the problem merely based on a few thoughts made off the cuff. One ought to check carefully whether certain components in our ecological system might perhaps still prove vulnerable with respect to ultraviolet radiation. He considers this to be rather unlikely, however.

Besides biological effects, a depletion in the ozone content of the atmosphere is likely also to cause changes in the temperature distribution of the stratosphere. This would probably have consequences for weather and climate. For the time being however, there is still a wide margin of disagreement among meteorologists about the nature and magnitude of the changes that might be expected.[38]

5

Between the devil and the deep blue sea

If microorganisms were human beings

If microorganisms were human beings capable of realising the consequences of their activities, they probably would have been concerned about the possible aftermath and side-effects of photosynthesis. Thus when this newly invented technique came into use some four thousand million years ago, they might have arranged a hearing to discuss these environmental problems in public.

At such a meeting the representatives of the anaerobic-conservative environmentalists would certainly have made clear their serious concern about a distortion in the competition that was likely to result from the introduction of this new technique. They were not against a limited application of photosynthesis, they might have said, so long as the raw materials used, such as hydrogen sulphide, did not pollute the environment. An expansion of this technology however, such as to continually produce large amounts of extremely poisonous oxygen as a waste product on a permanent basis, had to be firmly rejected. Otherwise the waste from photosynthesis would eventually contaminate the whole of the environment.

The proponents of the blue-green algaic liberal progressive party might have replied that in view of the predicted growth rates, the future energy demands of the unicellular race could not be met by fermentation and decomposition of hydrogen sulphide alone. Thus there was no alternative to the use of cheap water as a raw material. There would not be any risk of polluting the environment with the waste products of photosynthesis, since the poisonous oxygen would immediately be sequestered by bivalent iron. Any concern that these iron reserves might soon be exhausted were unfounded because this substance would permanently be replenished in large amounts from space outside the hydrosphere.

The anaerobic-conservatives might then have objected that the processes of iron supply from space were still not fully understood. It would be imprudent to base the energy supply of large sections of the community on a process whose volume and reliability were not yet known. If suddenly, for whatever reason, the supply of bivalent ferrous iron were curtailed or even totally interrupted, the only choice would be between perishing through a shortage of energy or through oxygen poisoning.

The blue-green algae liberals would have rejected these objections as being defeatist, being without any real foundation – and they would have been wrong. But nobody, not even the most visionary supporters of the new technique, would have dared to predict then that this controversial poisonous waste product of photosynthesis would, in years to come, indeed after some two thousand million years, serve their own descendants for respiration. It was even less predictable that just this pollutant would pave the way for the 'conquest of space' outside the hydrosphere, and also for the evolution of higher beings that would eventually end the two thousand million years of world domination by blue-green algae.

The reader of this book should not misinterpret this as a subliminal allusion that the nuclear waste we add to our environment might eventually, after some two thousand million years, serve the respiration processes or act as a source of energy for future beings that might evolve from us as we have from the blue-green algae. I just want to point out that our environment is an extremely complicated system with many interwoven feedback mechanisms. This implies that it is extremely difficult to assess the possible consequences and side-effects of a newly introduced process, and that even the most essential aspect might be overlooked.

If primeval man had been capable of realising the consequences of his way of life there might have been a public hearing when agriculture was invented some ten thousand years ago about the possible environmental impact of this newly introduced technique. Then the representatives of the liberal hunters' party would have emphasised in passionate speeches that man, being different from the animals in that he walks upright, was not born to bend his back to work in the fields and to win his bread by the sweat of his brow.

And the adherents of a natural-biological way of life would have pointed out that living on plants that can only thrive when sprinkled with excrements from animals was not only unhealthy but disgusting.

However the representatives of the agrarian prosperity party would have easily pushed these objections to one side. The pleasant taste of wild strawberries and quails' eggs is all very well, they would have said, but such naturally grown delicacies alone would never feed all the people if murder of newborns as a means of family planning were to be abolished. Moreover, the representatives of the hunters should not conceal the fact that due to permanent droughts and to excessive hunting quotas, the stock of game had already been reduced to such an extent that there was no alternative than to hunt each other. Without arguing about the gastronomical and digestive qualities of human meat, only the invention of agriculture would make possible a humane and also economic combination of war with adequate food supplies. Now the prisoner of war – after all, a human being walking upright, too – would no longer be used for improving the menu but rather for working in the fields. Therby he would produce more food than needed for his own sustenance, and also more than would be available if his body were to be eaten. One could well foresee that in the future wars would no longer be necessary in order to fill the flesh-pots with prisoners or to occupy a certain hunting ground but solely to save starving people from their sad existence of cannibalism and make them into sterling and well-fed slaves in a flourishing agrarian community.

The advocates of the newly introduced agriculture might have argued this way, and they would have been right in the end. But neither they nor their sternest critics then would have been able to foresee that with the invention of agriculture a chain-reaction had started, whereby solving one problem again and again created a newer and more serious one. They could not foresee that agriculture would initiate a world population explosion, an increase by a factor of one thousand within ten thousand years; that the attempt to feed this steadily increasing number of hungry people would again and again lead to new and more massive impacts on the environment that would change the face of the Earth beyond recognition – and that, nevertheless, misery and poverty would not be eliminated.

If we, the present humanity, were genuinely humane and not only concerned with the consequences of our own activities, but also felt compassion for the needy, we would find the misery of the Third World unbearable. We would do our utmost to help all the people on Earth ensure a somewhat dignified existence as soon as possible. Yet were we to try, based on the technical progress that has blessed us, the inhabitants of the industrialised countries, with all our considerable wealth (at least compared to the developing countries), we were likely to trade the devil we know for the devil we don't. We would most likely destroy our environment to such an extent that a somewhat dignified existence for *all* the people on Earth might well be jeopardised.

This is the trap in which we have caught ourselves, and from which we must now extradite ourselves as did Baron Münchhausen, with the top of our own heads. This is the very reason why a satisfying solution to environmental problems, as they are outlined in this book, is so difficult.

It is indeed so 'obvious' that to avoid the risk of an increase of carbon dioxide in the atmosphere we 'simply' ought to stop the growth of energy consumption, thereby also diminishing the pollution by oxides of nitrogen. This does not necessarily imply that our standard of living could not rise any more. We are used to wasting so much energy which could be better utilised. To carry through such a programme, a lot of stiff-necked opposition would certainly have to be overcome. In the Federal Republic of Germany for instance, it was impossible to maintain the speed limit for the motorways introduced during the oil crisis, although this measure unlike any other would have saved energy, money and reduced the casualties from accidents. In the USA, President Carter's proposal to impose higher taxes on automobiles with excessive petrol consumption proved unappealing in Congress. But whatever the difficulties are in reality, one can at least imagine that in theory highly industrialised countries could do without further appreciable increases in energy consumption.

In countries like India, however, *less than one tenth* of the energy is being used per head of population compared to that in the

Federal Republic of Germany and *less than one thirtieth* as compared to the USA – and this includes the burning of wood and dung which accounts for *half* (!) the total energy demands of that country. It is completely unrealistic to impose a zero growth rate with respect to energy in such a country, or merely even a zero growth per capita, which would still result in a doubling of the total energy consumption within about twenty-five years. In most countries of the Third World, together comprising three quarters of the world population, the situation is similar to that found in India. Without a *many-fold expansion* in the world energy budget it is virtually impossible to provide even the most primitive conditions that would ensure a somewhat dignified existence for all these nations.

If this increase in energy production is based mainly on coal and other fossil fuels, however, we are likely to face all the consequences that have been elucidated in this book: a drastic increase in the atmospheric carbon dioxide content and associated changes in the climate that in turn would hit the developing countries most severely, a growth in the abundance of the oxides of nitrogen, ozone depletion, and enhancement of ultraviolet radiation. Therefore, should we drive away the devils that appear from burning fossil fuels, with the super-devil of nuclear energy?

After the next ice age

Plutonium, being produced in considerable amounts in every uranium reactor, is certainly no nuclear waste. Rather, it is a highly desired product for both military and civilian purposes. In the USA, the first nuclear reactors went into operation during World War II to produce the plutonium required for the bomb that destroyed the Japanese city of Nagasaki. The first major 'peaceful' nuclear reactor, put into service at Calder Hall, England, actually is a multi-purpose installation producing both electricity for the public and plutonium for military purposes. The original design of this reactor even had to be altered in a way that was disadvantageous for the production of electricity, because the General Staff had demanded higher production of plutonium.[1] And we are not going to spend time in this book discussing whether all the many countries that today want to have nuclear power plants are merely interested in a peaceful utilization of atomic energy.

Yet in fact plutonium is not only used as a raw-material for nuclear weapons, but also as nuclear fuel for reactors. For both these applications, it has to be separated from the spent fuel elements, in fully automatic processing plants. With present-day reprocessing techniques over ninety-eight per cent of the plutonium can be extracted, the rest remaining with the nuclear waste.

In light-water reactors, the most common reactor type used in the USA, the Federal Republic of Germany, and most other 'nuclear' countries, the annual plutonium production amounts to 0.2 kilograms per megawatt capacity. Modern nuclear power plants are of the order of a thousand megawatt and more, and this corresponds to a plutonium yield of at least two hundred kilograms per year. The average lifetime of such a reactor can be assumed to be about thirty years. During this time it will produce about six tonnes of plutonium, over ninety-eight per cent of which can be extracted in the reprocessing plant. The rest, 60 to 120 kilograms, remains with the nuclear waste and has to be stored with it in an absolutely safe place until the radioactivity had died away.

With respect to geological time-scales, plutonium is a short-lived radioactive element. For plutonium 239, the most abundant isotope, the half-life is 'only' twenty-four thousand years. This half-life is the time required for half of a certain amount of plutonium to be reconverted into uranium 235, and hence half of the initial radioactivity will have died away in this time. The other isotopes have shorter half-lives; plutonium 240 for instance has a half-life of 6.6 thousand years. For every ton of plutonium 239 produced today, only one millionth of a gram will be left after one million years (40 half-life periods). Thus if the Earth were to have contained plutonium at the time of its formation, this radioactive element would have long since disappeared.

According to normal human scales however, twenty-four thousand years is a very very long time span. The first members of our own species appear some fifty thousand years ago. If plutonium 239 were to have been produced at that time, about twenty-five per cent of its initial radioactivity would still exist today. The pyramids in Egypt are more than four thousand years old. If their builders were to have stored plutonium in the pharaoh's vaults, its radioactivity would have only decreased to the present by about ten per cent.

Some scientists predict that at present we are on the eve of a new ice age. If this prognosis is right and if no excessive warming of the environment due to human activites were to counteract this development, the German depot for nuclear waste, planned to be established in a salt mine in the lowlands of North Germany, might perhaps be covered by ice some thousand years from now. According to the past history of ice ages, this ice cover might well last for some forty thousand years. If then the ice were to withdraw enabling man to repopulate this area, the plutonium, about a hundred kilograms per reactor, that is to be stored within the next few decades, dispersed in the nuclear waste, will still possess about twenty-five per cent of its original radioactivity. Thus, after the next ice age it will still be a lethal substance. But who will know exactly where these terrible materials had been stored, before the region was covered by ice?

'Plutonium is something terribly ultimate,' says E. Broda. 'Once it has entered the environment one cannot get rid of it. Do we really have the right to thrust on several thousand future generations the responsibility of standing guard over this toxic substance?'

Professor Broda is an expert on nuclear energy whose qualifications cannot be doubted. 'Since the foggy winter evening in 1941 when I entered the "old" part of the Cavendish Laboratory to take part in the British project for the utilisation of nuclear energy, I have in one way or the other been concerned with this field,' he told me. He built up a practical course in nuclear chemistry at the University of Vienna, worked as an official expert on nuclear energy, has been a member of the official Commission on Radiation Protection of the Republic of Austria for more than fifteen years; he was official reporter on questions of nuclear energy for the World Power Conference, served on the editorial board of several journals on nuclear science, wrote many scientific and popular books and articles on these questions and taught in Asia and Africa for the International Atomic Energy Agency. 'Like many scientists of my generation,' he says, 'I began as an enthusiast for the peaceful application of nuclear energy and I still do not consider myself as an absolute opponent. However, gradually I developed a critical attitude, and now my scepticism is quite pronounced.'

As an expert fully aware of the problems he does not deny that very grave difficulties would arise for the world energy supply, if the production of nuclear energy were stopped immediately. He is above all concerned about the long-term prospects: the possibility that in the next century the growing energy demand will be mostly met by nuclear power and in particular by the so-called plutonium breeders – reactors producing more new plutonium than is used as nuclear fuel.

'Even assuming that the problems in the focus of discussion today,' he continued, 'such as reactor accidents, discharge of radioactive material during routine working, cooling water problems and so on, were all solved to our satisfaction, which unfortunately is not at all realistic, such an expansion of nuclear energy production would create new problems at an even higher level, that appear to me to be *virtually insoluble*. Therefore, I believe that precautions should be taken in time and that a so-called "plutonium-economy", where the energy production and hence the economy are largely based on plutonium, must not be allowed to evolve at all.'[2]

This opinion, by the way, is in full agreement with the resolutions of the highly respected Pugwash conferences, where the majority of scientists from East and West have pleaded 'that the extent and duration of civilisation's dependence on nuclear fission should be kept to a minimum'.[3]

It is particularly worrying that with the further expansion of nuclear technology, more and more countries get access to nuclear weapons. Since each country will strive to be independent of the world uranium market and to reduce the risk of accidents and hence avoid the long-haul transportation of the highly radioactive spent fuel elements, more and more countries or groups of countries will inevitably establish their own facilities for the enrichment of uranium and for reprocessing the spent reactor fuel. Within the shortest of time-spans, however, every one of these nuclear plants – even if this had not been its original purpose – can be converted into a factory producing raw material for nuclear bombs. It is no wonder that a study group commissioned by the

American Ford Foundation of highly qualified scientists from different fields – including physicists and biologists, political scientists, jurists and economists – came to the sensational conclusion, published in their report in early 1977, that the USA should if necessary even do without the 'limited economic advantage' of developing nuclear energy, if this could prevent a further proliferation of nuclear weapons.[4]

The energy policy proclaimed by President Carter must be seen first of all as an attempt to tackle this problem in the twelfth hour – or is it the thirteenth already? According to Broda, this is at least a first step in a right direction. It remains to be seen, however, to what extent the President will succeed in implementing his programme against considerable opposition both at home and abroad.

So far the EEC countries have met the ideas of the American President with great reluctance. In particular the Federal Republic of Germany and France declared that they could not do without developing plutonium breeders, since they have neither the large reserves of fossil fuels nor the cheap uranium like the USA. Moreover, they were not willing to be excluded from the field of exporting nuclear technology, including reprocessing plants for instance, such as those sold to Brazil. Despite secret negotiations that have already lasted for years, all attempts to reach an agreement between the reactor-exporting countries, to put an end to the possible proliferation of nuclear weapons, so far have failed.

Besides the risk of more and more countries obtaining access to nuclear weapons, a worldwide proliferation of nuclear technology would also imply other serious long-term problems. The necessity to protect more and more nuclear reactors and other nuclear installation against sabotage and terrorists, gangsters and lunatics, could cause a political development towards a police state.[5] And finally, plutonium is not only a raw material for producing bombs, but, moreover, as Broda puts it, 'it is a really devilish substance from the point of view of radiation protection; of all materials produced in larger quantities, it is by far the most toxic'. This chemical element, produced in every uranium reactor, is a very strong so-called alpha-emitter. It is relatively easy to protect oneself against such radiation when it originates from an external source. It has a disastrous effect, however, if it becomes incorporated in the body. Due to the relatively long half-life, the intensity of the radiation virtually does not decrease throughout a human lifetime. Consequently the full radiation dose is maintained as long as the element remains in the body. If dust containing plutonium is inhaled and if only a few of such dust particles settle on the lung tissue, lung cancer is likely to develop. If plutonium is injected into the blood stream or if it enters the body with the food, it is mostly deposited in the bone marrow and the liver, and both of these organs are particularly sensitive to radiation. For this reason, the upper limit that can be tolerated for persons working with plutonium was fixed by the International Commission on Radiological Protection at the extremely low value of 0.6 millionth of a gram. This corresponds to a dose rate that is five times smaller than that for radium.

If the world energy supply were based mainly upon nuclear fission, many thousands of tons of plutonium would have to be produced, transported and processed every year. Occasional accidents would be inevitable, whereby some of this material would escape into the environment. There would be practically no way of cleaning contaminated areas. One would simply have to close these for some hundred thousand years and take precautions that no plutonium containing dust or contaminated water from the prohibited areas gets into neighbouring inhabited regions. But if this 'plutonium economy' were to last for many decades or even centuries, as intended by its supporters, more and more regions would eventually be contaminated, and a global increase in the plutonium burden could hardly be prevented. With regard to the time-scales of human history, this would be a completely irreversible process. As mentioned already, twenty-four thousand years are required for the radioactivity of plutonium to decrease by half.

It is far beyond the scope of this book to present in some complete way the arguments for and against nuclear energy. In connection with the subject dealt with here, it is mostly the long-range problems that are of interest – the question as to whether or not it might be possible and reasonable to counteract the threatening increase of atmospheric carbon dioxide by a rapid and worldwide deployment of nuclear reactors. In this section we have touched on some of the reasons why many competent and responsible scientists reject such a development. But do we have another alternative?

A third alternative?

Even if it were true that we only have the choice between a plutonium economy and burning so much fuel that the carbon dioxide increase will cause grave climatic changes, it is still an open question which of these two evils would be the lesser one.

When asked by the German climatologist Hermann Flohn whether I would consider salt lakes in the vicinity of Vienna as 'acceptable' (see page 103), I replied that actually there are salt lakes only fifty kilometres from Vienna near the Hungarian border *now*, and that nevertheless people are living there quite decently and not like Arab nomads. Of course, a gradual change-over to a Pannonian (Hungarian) climate, or even worse, in the Austrian plains would be a matter of grave concern indeed. Yet, as I see it, it probably would still be a lesser evil than a slow but sure increase in plutonium contamination.

Opinions diverge widely on this question. To Flohn, it seems that the risk of liberally using fossil fuel is 'at least as grave' as the risks from nuclear power stations which, he thinks, are 'technologically minimiseable'.[6] His American colleague Stephen Schneider, on the other hand, though continually urging the IIASA workshop not to underestimate climatic dangers even if they as yet cannot be spelled out in detail, told me that he is 'scared to death' by the prospect of an unregulated plutonium economy.

It should be emphasised that such views must be considered as personal opinions rather than as judgements of scientific experts. For even discounting all uncertainties involved in making a prognosis, there simply is no 'scientific' method of objectively comparing the consequences of plutonium contamination and its increased incidence of cancer, leukemia, hereditary diseases etc., with the climatic consequences of carbon dioxide increases and the resulting economic hardships, harvest failures, famine etc. The comparison cannot be but a personal value judgement and values cannot be placed in a true scientific category.

But is there *really* no other choice than either a plutonium economy or a dangerous build-up of atmospheric carbon dioxide, if we want to have 'enough' energy? After studying these questions for several years I am convinced now that we are not faced with such a dire choice. There *are* ways out of the dilemma. We can have a world with an acceptable level of energy supply in an environment neither contaminated with plutonium nor threatened by climatic catastrophes and we can have it at a reasonable price compared to the price of other alternatives in terms of money and 'quality of life'.

In order to find a way out we must deal with four questions:

1. How is the demand for energy likely to develop within the next hundred years?
2. What is the real scope *and urgency* of the carbon dioxide problem?
3. Are there alternative ways to provide energy without emissions of either plutonium or carbon dioxide and *when will they be available*?
4. Is it possible *to bridge the time gap* until such alternatives can be widely deployed without an intolerable increase in the atmospheric carbon dioxide burden?

As far as the first question is concerned, one point should be made clear right at the beginning. A 'solution' to the carbon dioxide problem by perpetuating the underconsumption in the developing countries is certainly not acceptable. 'A worldwide renunciation of growth of energy production would lead to social catastrophes, certainly as unbearable as the climatic catastrophes that would most likely otherwise occur,' writes Ulrich Hampicke of Essen University. 'Even an Arctic Sea with no ice would be a less serious consequence than a World War between the poor and the wealthy that is likely to result from fights for the redistribution of resources.'[7]

In the countries of the Third World it is virtually impossible to establish living conditions that are somewhat dignified without allowing a massive growth in energy consumption. In the highly industrialised countries however, there is no irrefutable law saying that economic growth, and even more so raising the quality of life, must be closely coupled to steady increase in energy consumption.[8] In the USA, the energy consumption per capita amounts to about twice that in Sweden, while the standard of living, however, is about the same in both countries and there is certainly even less genuine poverty in the Swedish welfare state. Thus, earlier predictions for the world energy demand, based upon the *waste of energy* typical for the USA, must be considered exorbitant today.

As late as 1972 Alvin M. Weinberg, the highly respected American nuclear physicist, published with R. P. Hammond as coauthor a study[9] based on the fact that the standard of living, even in the wealthy USA, is unsatisfactory for large groups of the population. In their view a saturation in the demand for energy would not be reached until the consumption per capita was twice as high as the 1970 US rate. This would mean a consumption of about twenty tonnes coal equivalent per capita and year. With a world population predicted then to reach fifteen thousand million by the middle of the next century, this would raise the world energy consumption to three hundred thousand million tonnes coal equivalent per year – *about thirty-three times as much as that in* 1974! The energy project group of IIASA came up with not much lower figures at that time.

At the Dahlem conference in 1976, Weinberg presented a paper coauthored by R. M. Rotty, that was already considerably more modest. It predicted that the world energy consumption would level off at about eighty thousand million tonnes coal equivalent by the middle of the next century with no further increase.[10] For a world population of fifteen thousand million, this would correspond to a global average similar to the present consumption per capita in Western Europe.

In the meantime the forecasts for the world population growth rate have also been corrected to give smaller values. At present 'only' ten to twelve rather than fifteen thousand million people are expected to populate the world by the middle of the next century. Based on this figure, and with more thought being given to the ways to save energy in the highly industrialised countries, a new prognosis was made by IIASA. For the year 2030 it comes up with a world population of about eight thousand million and a consumption of energy amounting to thirty-five thousand million

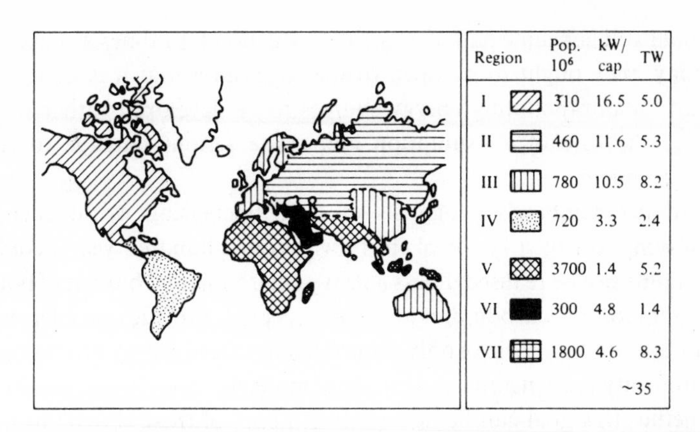

Region		Pop. 10^6	kW/cap	TW
I		310	16.5	5.0
II		460	11.6	5.3
III		780	10.5	8.2
IV		720	3.3	2.4
V		3700	1.4	5.2
VI		300	4.8	1.4
VII		1800	4.6	8.3
				~35

Figure 18 IIASA scenario for the world energy demand in the year 2030 assuming a world population of eight thousand million. kW/cap is roughly equal to tonnes of coal equivalent per head, TW (terrawatt), meaning terrawatt-years/year, is about equal to thousand million tonnes of coal equivalent per year. (From Häfele.)

tonnes coal equivalent – about four times as much as in 1974.[11] This would certainly not reflect the highest level yet as the energy demand is likely to further increase after 2030, but it is after all an important and apparently realistic guideline.

Looking at the regional forecasts in this scenario (see figure 18) it becomes quite obvious that the differences between poor and rich countries will still be frighteningly large, even fifty years from now. However according to this prognosis they will at least become smaller. For a country like India, such scenario predicts an increase in the consumption per capita by a factor of 3.5 compared with the present, corresponding to an increase of the total consumption by a factor of ten if the growth in the population is taken into account. An even much more rapid development would certainly be desirable, but it would probably be unrealistic to assume that this could be achieved.

For the global average this scenario would imply a doubling of the consumption per capita in the next fifty years. This is no final goal, but it would certainly not be intolerable, in particular for the highly industrialised countries, to accept that this rate of development should not be exceeded. Since these countries at present consume seventy-five per cent of the world energy production, and would still consume half of a world production four times as large as today, they ought to be open to the suggestion that it is better to wait for non-polluting energy sources to be developed rather than force the rate of consumption and cause serious damage to the environment.

Alvin Weinberg's prognosis of 1972 (increasing world energy consumption by a factor of thirty within one hundred years) could certainly not be realised unless a heavy environmental burden of both plutonium *and* carbon dioxide were accepted. For the considerably more modest scenario of only a fourfold increase in energy production within fifty years, it appears not to be completely unrealistic to consider whether this goal can be achieved with *both* of these threats to the environment being avoided.

Is the carbon dioxide issue being overemphasised?

One would guess that the assessment of future trends in the atmospheric carbon dioxide concentration is exclusively the field of meteorologists, geophysicists, and possibly also ecologists, but certainly not of nuclear scientists. In fact however, the author of one of the aforementioned forecasts (see page 92) is K. E. Zimen, a nuclear chemist and former director of the Hahn-Meitner-Institute for nuclear research in Berlin. Similarly the problem of carbon dioxide increase is being studied at many other nuclear research establishments all over the world. The reasons for such excursions into an unacquainted field of research are quite obvious. 'We wanted to find out how acute the carbon dioxide problem really is,' Zimen told me at the Dahlem conference. 'That is to say, we wanted to know how fast and to what extent nuclear energy has to substitute fossil fuels.'

It is understandable that scientists involved with nuclear energy see the problem of carbon dioxide increase from this point of view. It is an indisputable fact that nuclear reactors, as opposed to conventional power plants, do not release carbon dioxide into the environment. In the discussion about nuclear energy this is of course brought forward by the reactor industry as an argument for their products. Thus, it is not surprising that General Electric Company, the world's largest producer of nuclear reactors, had two of their scientists participate in the IIASA workshop on the carbon dioxide problem.

Representatives of the West German Coal Producers' Association were present at the workshop too, and one of them, Dr G. Zimmermeyer, pointed out that reactor producers are very eager to draw attention to detrimental environmental effects of modes of energy production competing with them on the energy market – especially since the announcement of President Carter's energy programme based mainly on the use of coal for electricity generation.[12] For the coal mining industry, the continual discussions about carbon dioxide are, of course, not at all desirable. Thus Dr Zimmermeyer urged the participants in the workshop to present these problems to the public only in the 'most responsible manner', since otherwise the building of conventional power plants might in future encounter the same problems as the establishment of nuclear reactors is facing right now. Since General Electric, and most other reactor producing companies as well, also build conventional power plants, he may hope that this side will show a certain understanding for his arguments.

In the free enterprise system everyone has the right to advertise his products and thereby refer to those scientific findings that favour his system. Nobody would expect an advertising manager to point out disadvantages in his products just for the sake of scientific objectivity. In the same way one does not expect a lawyer, in his speech for the defence, to present material incriminating his client just for the sake of objectivity. But the complicated character of modern industries such as nuclear reactor production often implies that it is only qualified scientists, rather than advertising agents, who can disperse expert information.

This is a general problem of modern society, as J. Ravetz, an English historian on science, pointed out. Today most experts qualified in the field of natural sciences do not work independently, striving for scientific knowledge, but rather as experts employed by some company or authority. It is part of their professional duties to adequately safeguard their employer's interests. Thus, says Ravetz, one would oblige them as well as the public, if one were to clearly allow them a role in society comparable to that of the lawyers. But from the tradition of his profession, his vocational training as well as the role society expects him to play, a scientist is supposed to commit himself to scientific objectivity, even if in an argument he had undoubtedly acted on behalf of the interests of one party. This is the reason for a great deal of confusion.[13]

It is unreasonable to expect a reactor specialist whose future career depends on more reactors being built to give an entirely objective judgement, free from personal considerations, as to whether or not nuclear power is desirable. On the other hand he should not demand that the public consider him as an objective expert in the discussion on nuclear energy.

The more nuclear power plants are planned, the more companies and people will become interested in the expansion of nuclear energy for the sake of profit and/or their professional careers. For this reason alone, the anti-nuclear movements fulfill an important role in society, even if their arguments are sometimes exaggerated and not well founded. They act to counterbalance the nuclear lobby and force the public to face these problems seriously. They prevent us from blithely accepting a plutonium economy, that could be very difficult to change later. 'The anti-nuclear movements can take credit for making the public aware of the problems related to the construction of nuclear power plants,' said Walter Scheel, President of the Federal Republic of Germany.[14]

It is also through such public initiatives that the nuclear industry is often charged of exaggerating the carbon dioxide problem, but in general the nuclear opponents are not able to substantiate such allegations in detail. Yet, a careful study of all the available facts, as is attempted in this book, shows that most likely there will be a carbon dioxide problem after all some time in the next century but that its *urgency* is often greatly exaggerated. For if it is true that the amount of carbon dioxide liberated by the clearing of tropical forests is about as big as the fossil fuel source, and there seems to be ample evidence for this (see page 85), it is only of minor importance to know exactly where most of this carbon dioxide goes now. The tropical forest source will fade out anyway within twenty or thirty years, because there will not be any tropical forests left by then that are worth felling (unless the wanton destruction is stopped before then). Thus the sinks, wherever they may be, in the oceans, in other parts of the biosphere, or somewhere else, will be able to accommodate much more fossil fuel carbon dioxide thereafter and the atmospheric burden will not increase as dramatically as predicted by the geophysicists' models. This will give some breathing space for the development of alternative energy systems.

It is worth noting that I heard such views from participants of the Dahlem conference too. One of them – he asked me not to mention his name, but he certainly has no connections whatsoever with the coal mining or other fossil fuel industry – told me rather emotionally: 'This is all just only grist to the mill of the reactor industry. One still does not know enough about the actual scope of the carbon dioxide risk. One is searching for the "missing sink", but there is neither agreement about how much carbon dioxide is really being liberated, nor where it is going to. The predictions about the future rate of carbon dioxide increase in the atmosphere are hardly little more than guesswork, and the extent of the possible climatic impact is under debate.'

'But if we wait until all these questions will have been answered with some reliability, it may already be too late to draw practical conclusions,' I objected. 'Change-overs in the energy sector usually require decades.'

'If there really is a carbon dioxide problem, it will not be urgent before another fifty years,' he replied. 'This cannot be used as an argument to stop building conventional power plants now and instead establish only nuclear reactors.'

'Certainly not, but it should be a reason to search for other alternatives, right now and with much more effort than in the past.'

'Sure, I agree. What we need is a crash programme, intense efforts to develop techniques for utilising solar energy.'

Always just 'one horse, one rabbit'?

An old but still good joke tells the story of a traveller who had ordered rabbit pie in a little Balkan restaurant. When the dish was served its taste appeared strange to him and not just rabbit-like. He called for the innkeeper.

'What did you serve me there?'

'Rabbit pie, sir.'

'Don't give me that. This is horsemeat, but no rabbit!'

At first the host tried to deny this, but since he could not pacify his guest he eventually admitted in a friendly way that he had 'stretched' the rabbit, which was in compliance with the laws of his country.

'What do you mean by stretched?' said the annoyed guest. 'This is pure horsemeat!'

'No, sir,' reassured the innkeeper, 'this is fifty-fifty, exactly as the law requires – one horse, one rabbit.'

This joke was told during the Dahlem conference, when Alvin Weinberg pleaded that *all* alternative methods for energy production should be pursued as intensively as possible. In reply he was told that in the USA, and not only there, research funds were in fact distributed among nuclear energy and other alternative techniques such as solar energy, according to the 'one horse, one rabbit' principle. In 1975, the American Energy Research and Development Agency (ERDA) spent 1.2 thousand million dollars on projects related to the utilisation of nuclear energy, but only 80 million dollars on solar energy research, although this was real progress when compared to the years before. In 1974 solar energy research received 14 million dollars from the ERDA, while in 1973 only 4 million dollars.

In their paper on alternative long-range energy strategies presented to the Dahlem conference, Weinberg and Rotty, as mentioned before, had assumed a consumption ceiling of eighty thousand million tonnes coal equivalent.[15] In view of the increase in atmospheric carbon dioxide, they wrote, coal could certainly not be the main supplier of energy. Moreover, such a high consumption rate would soon exhaust the global reserves. Those possible alternatives that are relatively cheap and technically feasible such as water power, wind energy, geothermal energy from geysers and underground steam, solar collectors for heating of water, are limited by natural conditions and can provide only a relatively small contribution to the enormous global energy demand. Among the alternatives that could provide really large amounts of energy, nuclear fusion ('taming of the hydrogen bomb') and the utilisation of geothermal energy of deep dry rocks have not yet been proved to be technically feasible. For nuclear fission and in particular the plutonium breeders, there are the aforementioned reservations that they might be too dangerous; and solar energy power plants, according to Weinberg and Rotty, might probably be too expensive. Personally, he thought, explained Weinberg in the discussion, that the risk from the plutonium breeder could be reduced to a tolerable level and that this solution would certainly be the best one under the given circumstances. But since this opinion would not be generally shared, the only reasonable course for the time being would be to pursue all the alternatives as intensively as possible.

For the fact that the funding of solar energy research had been so low for so many years, two explanations were offered in the corridors of the Dahlem conference, both of them likely to be right. The first one suggests that the oil and reactor industry, interrelated in various ways, brought every influence to bear in order to slow down the development of alternative energy sources. This holds true in particular for the easily decentralisable solar energy that would reduce the control of some highly centralised corporations over important branches of energy supply. The second explanation says that a new field of research like the ultilisation of solar energy can only be developed step by step and that more funding cannot be spent usefully than at present. From a merely commercial point of view this is probably right. A crash programme, i.e. a forced development, implies necessarily that a gain of time has to be bought at the price of higher expenditures. Such a policy will not prevail unless the public and the authorities that provide the funds are convinced that nuclear fission is merely a *temporary solution* for providing energy.

President Carter's energy programme tries to adopt a middle course on this question. He certainly wants to avoid a plutonium economy (fast breeders, commercial reprocessing plants for nuclear fuel), but not do without light-water reactors. Accordingly, the nuclear energy branch is still the focal point of expenditure in US energy research. The funding for solar energy and other alternative sources of energy has certainly been greatly increased, but the view that these research programs should rank *highest* in the priority list has not yet gained acceptance, and moreover the President's programme is encountering great difficulties in Congress.

As long as the technology for the use of solar energy remains in an early state of development however, it is difficult to judge whether its widespread application would really be as expensive as Weinberg and Rotty maintain. In all events, since the oil crisis, the prices for fossil fuels, uranium and nuclear reactors have all increased much faster than the inflation rate with further increases likely, whereas for many components for solar energy, in particular photovoltaic cells, there is a marked trend towards lower prices. Moreover, the calculations of profits from nuclear reactors themselves are highly questionable since they contain numerous open and hidden subsidies without which reactors could not compete with traditional power plants at all. Enriched uranium for light-water reactors is still being supplied by the American Governmental Industries (originally established for military purposes) at prime costs. In most countries where nuclear reactors operate, the liability insurance for the nuclear reactor plants is limited to levels far below the possible damage likely to be caused by a major accident, and even for part of this sum, the state is held liable for partial failure. The costs of decommissioning a worn-out nuclear reactor have not been established as yet and no company operating nuclear plants has the means to guarantee a safe storage of the nuclear waste for several hundred thousand years. This burden is being imposed, without asking them, on future generations.

Speaking of solar energy, Weinberg raised the question of energy storage during periods of bad weather, and its costs. But basically this problem exists for *every* power plant that due to technical or other reasons can fail for a few days. It is in particular the nuclear reactors that, because of their technical peculiarities, have to be shut down from time to time, for weeks and even months as a rule, without causing a breakdown of electric supply. Thus every power supply grid has to carry a reserve capacity to cater for such events. Even if solar energy were to become the major source of energy production this would certainly not imply that all hydroelectric and caloric power stations would be abolished, and moreover it is unlikely that all the solar energy power plants of a large power supply system would simultaneously experience bad weather for several days.

A particular way of utilising solar energy, the technical feasibility of which still has to be proven however, is envisaged by E. Broda.

It involves the production of hydrogen through the photolytic dissociation of water using sunlight. Basically, this process is analogous to photosynthesis in green plants, but Broda's idea is not simply copying the plants but rather learning from them.[16] Research in this field would be very cheap as in the initial phase only a few qualified scientists would need to be paid to study this problem in detail. Only when the results of such studies are available will one be able to judge whether this idea can be realised at reasonable costs. The hydrogen produced in this process could be utilised in a similar way as natural gas, but with the important advantage that its burning produces no carbon dioxide but merely water vapour.

'Wouldn't the hot hydrogen flame also cause oxides of nitrogen to be formed?' I asked.

'Certainly in a flame, but this can be avoided if the hydrogen is not burnt but rather utilised in fuel cells, and their development is proceeding rapidly,' was the answer. This process, also called 'cold combustion', is somewhat the reverse of electrolysis. In the fuel cell, hydrogen and oxygen react forming water. There is essentially no heat produced, and the energy liberated through this process is available as electricity.

The main problem of solar energy today is the time factor. Theoretically the Sun can contribute considerably to domestic heating, but houses with effective heat insulation have to be built first. Even if appropriate construction regulations were enacted, it would require decades until a good percentage of all the houses were to fit this standard. Solar collectors for heating of water are commercially available and comparatively widely utilised already in several countries, like Israel, Japan, and Australia. In most Mediterranean countries however, they are as yet very sparse.

In 1977, solar energy covered merely one twohundredth of a per cent of the total energy demand in the USA. According to a feasibility study made by the Office of Technology Assessment, almost ten per cent of the total US energy requirements might be met by decentralised exploitation of solar energy.[17] This figure includes electricity produced with photovoltaic cells likely to be available at acceptable costs in near future. But to reach this goal of ten per cent before the year 2000, the exploitation of solar energy would have to expand by about fifty per cent every year. Even with the most massive support programme imaginable, the realisation of such an intensive growth rate must be doubted.

In several countries, experiments with prototypes of large solar power plants have begun. It will be several years however, before sufficient experience is to hand, and *decades* before such novel power plants can be deployed everywhere in the world. But how shall the energy requirements be met *until that time*, if on the one hand one attempts to do without fast breeders and to limit the total of light-water reactors, while on the other hand, the carbon dioxide burden of the atmosphere is not to exceed a tolerable level? Is it possible to provide the necessary breathing-space?

A strategy to buy time

A global transition from one main energy source to another one requires about fifty years, declared Wolf Häfele, Chairman of the IIASA energy project, in his opening address to the carbon dioxide workshop at Baden. If present trends were to persist, the carbon dioxide burden of the atmosphere might have reached a dangerous level by the year 2030. Thus, in Häfele's view, it is of the utmost urgency to start developing alternatives *right now*.[18]

It is remarkable that the scientists participating in the IIASA conference *did not support* Häfele's appeal for immediate action. Rather, the working group on energy strategies, under the chairmanship of R. M. Rotty of Oak Ridge Associated Universities, declared that the uncertainties in nearly every aspect of the carbon dioxide issue are too great yet to justify a major change in energy policies. 'Mankind needs *and can afford* a time window of between five and ten years for vigorous research to narrow these uncertainties,' they declared. In the meanwhile to 'maintain flexibility is the only prudent course of action'.[19]

The report of the working group stressed 'that it is *premature* to implement at this time, policy measures which require the reduction in the use of coal and other fossil fuels ... On the other hand, policies to emphasise the use of coal because of its abundance in preference to non-fossil energy supply systems are equally unjustified. Such policy decisions can become difficult and very costly to reverse because of the evolving technologies required in each energy supply system.' It would be highly desirable, the report stated, to devise energy supply systems producing only little carbon dioxide or none at all and/or to find ways and means to mitigate undesirable environmental effects.

Besides other alternatives, either limited by nature or probably not going to be available soon enough for global deployment (see page 152) the Baden workshop also thoroughly discussed the idea, suggested by the IIASA, that methanol be the main carrier of energy for the next century.[20] Methanol would be advantageous because this liquid fuel can be produced from coal through a well-tried process. It could be used without requiring costly changes in the infra-structure and it could, at least for the time being, be added to the oil products, and gradually substitute these as natural oil resources become exhausted. Since only half the energy produced by combustion of methanol originates from the oxidation of carbon, the other half coming from the oxidation of hydrogen, the carbon dioxide problem would therefore be 'cut by half', as Häfele pointed out. His concept is that fast breeders or huge solar energy installations should be utilised to provide the considerable amounts of energy required to produce this easy-to-use liquid fuel. (From the fact that the nuclear physicist Häfele, known as an ardent supporter of the fast breeder, mentions solar energy at least as an alternative, one can guess that the numerous energy debates of the last few years have not been entirely in vain.)

According to the IIASA concept, these breeders and the necessary plants for reprocessing the nuclear fuel should be established under UN control, on remote natural or man-made islands in the ocean. Thus no long-range transportation of dangerous radioactive material would be necessary, and in the event of an accident, no one besides the operating personnel would be exposed to danger. By means of such highly centralised 'island projects', the risks of a gradual global plutonium pollution, of 'diverting' plutonium for military purposes or terrorism, and of sabotage could not be completely excluded, but they would certainly be much lower than in the case of fast breeders being installed in many countries for their immediate energy supply.

A second project investigated by IIASA deals with the possibility

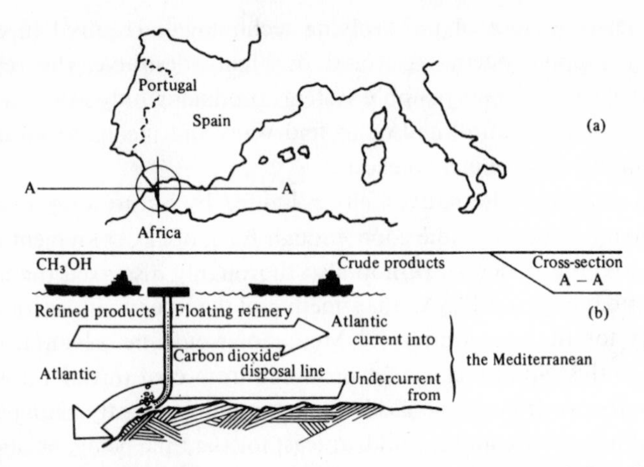

Figure 19 Scheme for removing carbon dioxide by releasing it directly into the deep ocean waters from a floating plant working in the strait of Gibraltar. The plant might produce methanol, hydrogen or electric power using fossil fuel as an energy source. (a) Location. (b) Scheme. (From Marchetti, *Climatic Change*.)

of dissolving carbon dioxide exhausted from conventional power plants and other major combustion installations in the oceans. Instead of letting the carbon dioxide pollute the atmosphere, it would be separated from the exhaust gases and led into the deep ocean waters which, being undersaturated in carbon dioxide, have a virtually unlimited capacity to dissolve this gas.[21] One would have to operate such installations at places where the ocean currents themselves could carry the dissolved gas into the deep layers, such as at the Strait of Gibraltar for instance, where a current into the deep Atlantic is effective. Based on such conditions, the costs of the project would not be prohibitively high.

A further possibility to remove carbon dioxide from the atmosphere, 'the planting of trees', is only cursorily mentioned in the IIASA report, and although it gives me some personal satisfaction that these few words were added on my suggestion, I cannot help feeling that this alternative is underemphasised. Throughout the workshop I was under the impression that most of the energy experts, engineers and also many of the physicochemical and geophysical scientists taking part were preoccupied with searching for technological solutions while underestimating the great possibilities of a reasonable forest policy.

The Dahlem conference with its stronger representation of ecologists and other bioscientists was much more outspoken on this point. One of its recommendations was 'to study the feasibility of massive reforestation and management of presently deforested areas and of improved wood economy'. A reversal of the trend in present deforestation, states one of the conference's reports, 'might conceivably provide, within fifteen to twenty years, an additional biospheric sink of the order of the present input ... Such an action, coupled with an extensive wood economy to replace other structural materials (cement, aluminium, plastics) which are energy intensive, may present multiple benefits at least for the period required for the development of more suitable energy alternatives, and postpone the pressure towards adopting the nuclear energy option.'[22]

Reforestation – a policy with many benefits

A worldwide large-scale project of reforestation would not be a technological solution. Rather it would be a good attempt to solve an environmental problem by a return to more natural conditions. *Just for this reason* reforestation would bring many other ecological benefits quite apart from mitigating the carbon dioxide problem.

'Priority should, in any case, be given to measures like reforestation which are safe on environmental grounds and which have an economic impact relatively easy to assess in terms of manpower, agricultural land loss, etc.,' says B. Schefold, a German economist, in an appendix of the Dahlem report. 'In working out an international program of reforestation, one should take account of the economic and social needs of the nations most immediately concerned, of unemployment etc.'[23]

According to George Woodwell there can be no doubt that such a programme, coupled with one that preserves existing primary forest, would be effective in removing carbon dioxide from the atmosphere. In a letter to me he writes: 'The evidence we need is clear in the data of the carbon dioxide concentration in the air taken at Mauna Loa. Here the annual withdrawal of carbon dioxide from the air reduces the concentration by about 5 ppm. [See also page 4.] It is of course replaced during the winter when respiration dominates over photosynthesis. I would guess that we would make a very substantial step toward improving the human lot by simply adjusting our activities to the point where no further deforestation occurs. If we could shift the pendulum in the direction of reforestation, we would be making a very large contribution towards the stabilisation of the land surface and other essential components of the biosphere.'

The sooner one would start the better, because trees can certainly be planted more easily in areas recently cleared, where some humus is still available, than in regions that are already completely eroded. But even mountains once covered by forests, that are now

Figure 20 Reforestation of Aleppo pine trees on nearly bare rocks in the Sibenik region, Dalmatia. The trees were planted at the turn of the century. Soil had to be carried in to enable planting. (Photograph: by courtesy of H. Leibundgut, Institute for Forestry, ETH Zürich.)

totally barren, can be reforested if proper care is taken. In Dalmatia, there are pine forests which have been planted virtually on bare rocks. Soil had to be carried in to enable planting. Figure 20 shows the result of such reforestation after fifty years. One can hope that a new humus layer will gradually accumulate in the protection of the trees enabling soil vegetation to develop. Man certainly has to assist in this process, at least by preventing sheep and goats from grazing in the reforested areas.

Already for more than a century, there have been attempts at a partial reforestation in Dalmatia. But every holiday-maker who has toured that country knows how small and rare the forests are there and how vast the regions that are totally eroded or covered by Macchie bush. Large portions of the other Mediterranean countries are very similar.

'There is no question that much of Europe, especially the Mediterranean countries, supports a vegetation that is the product of heavy overgrazing, repeated burning, and constant harvest for firewood over millenia,' writes Woodwell in the abovementioned letter. The south European and even more the north African countries are too poor to accomplish a really large-scale afforestation program on their own. They are so poor, not least because in the totally deforested regions, a considerable proportion of their peasants, for hundreds or even thousands of years, have wrenched only paltry harvests from a stony, much too rapidly drying soil. The times when Sicily and northern Africa were the granaries of ancient Rome have long since passed. During antiquity the fertility of these countries had largely been destroyed by unrestrained exploitation of the soil. In many touching Italian movies, the poverty of the people that live in these regions has been documented. It is a poverty that has a feedback effect on the entire life of the country.

One example from the Provençal Alpes in southern France shows how much even a single man can achieve in such impoverished land. Between 1910 and 1945, Elzéard Bouffier, a peasant who had retreated as a shepherd to a remote and rather barren mountainous region after having lost his wife and son, planted hundreds of thousands of trees, spontaneously and without payment, thereby creating a forest that extended for some thirty kilometres. Jean Giono, a writer who had met Bouffier in 1913 on one of his excursions to the mountains and who later paid him several visits, describes the activities of this man. His report, although being not really a scientific document, clearly comprehends the essentials of Bouffier's achievements.[24]

Shortly after World War I, he reports, one could already notice the effects of this afforestation in the valley below: 'When I came down to the villages I saw water flowing in the river beds that had been dry from time immemorial. This was the greatest chain-reaction I have ever seen.' Finally even foresters became aware of this 'natural forest' that had, contrary to all previous experience, started growing 'on its own'. Delegations came to visit this 'miracle', and the forest was declared a sanctuary – but nobody knew about the man who had planted it.

In 1945 Giono paid his last visit to Bouffier, who was then 87, to see the results of his work in the valleys: 'I stopped at Vergons. In 1913 only three people had lived in this little hamlet of a dozen houses. They looked like semi-savages, with the physical and moral phenotype somewhat like prehistoric man. The abandoned houses had been overgrown with stinging nettles. The living conditions had appeared hopeless... Now the hamlet had twenty-eight inhabitants, among them four new households. The new houses, with fresh plaster, were surrounded by vegetable gardens. This had become a place where one would have liked to live in.'

And eight years later: 'Neat farmhouses rise up where I saw just ruins in 1913. The villages have gradually been rebuilt. In the streets one comes across men and women, boys and girls who know how to laugh... Thus more than ten thousand people owe their happiness to Elzéard Bouffier.'

It appears to be a symbolic tragedy that this peacefully grown prosperity did not last very long. In 1970 Giono wrote to the German translator of his report, who had wished to visit this region, 'Since that time everything has been changed in order to build silos for nuclear weapons, firing ranges and several oil reservoirs. There are only just a few firs and a few patches of beech and alder forests that have survived.'[25]

Elzéard Bouffier was active in a region that was still not completely eroded. He did not have to carry in soil, and sheep could graze there as well. Later he abandoned his flock, because the animals damaged his trees, and resorted to bee-keeping – since the planting of trees was, so to speak, his 'hobby', but not the profession he lived on.

His activity demanded no end of patience, devotion and persistence. According to Giono's report, for every hundred trees he had planted, only about ten kept growing.[26] But finally there was forest where only poor pasture had been before.

It is possible to make devastated and even totally eroded areas green again. It is possible to convert a desert into a prosperous land. Young enthusiastic people have repeatedly proven this in Israel.[27] But much effort is needed, and often it takes a long time until returns materialise.

A large-scale afforestation with all the means of modern science and technology could probably help ten thousands of villages in the Mediterranean countries to attain new prosperity. The goal should be the creation of almost natural forests that can finally survive and flourish on their own and that might sometime, in a hundred or even more years, yield a modest return of timber. Their real benefit, however, would be an indirect one; an improvement in the water conservation and other ecological conditions in the near environment, and on the global scale, a delay in the increase of carbon dioxide in the atmosphere.

The reforestation of the Mediterranean countries 'at our very doorstep' could be the long-range goal of a carefully planned initiative of all European countries, and would set an example for the whole world. This programme would need several decades to accomplish, but even with regard to the feedback with the atmosphere, it would nevertheless be worthwhile, since it would last into the next century when the carbon dioxide problem will really become urgent.

In the middle of this century, tropical rain forests covered an area of about seventeen million square kilometres, that is eleven per cent of the land surface of the globe. By the year 2000, if present trends continue, most of them will have been cleared. Moreover, there are other regions too, as for example Siberia or northern Canada, where evidently much more forest is being cut than is regrowing. If mankind is able to clear about ten per cent of the total land surface within a time-span not exceeding half a century, it should also be able to reforest about two per cent of the land surface within twenty or thirty years. Let us assume that reforestation could create about 1.5 million square kilometres of new temperate evergreen forest (mainly in areas where it had been destroyed centuries or even millenia ago, as in the Mediterranean countries) and restore about 1.5 million square kilometres of tropical forest (mainly destroyed within this century as for example on the slopes of the Himalayas). Taking the net primary productivity as stated in the table of Whittaker and Likens (see table 2 on pages 172–3) and assuming as a first guess that about sixty per cent of this amount would remain stored in the newly growing forests (the rest being reconverted into carbon dioxide by the respiration of microorganisms and animals), we can figure out that we could create in this way a sink for about five thousand million tonnes of carbon dioxide annually (1.3 thousand million tonnes of carbon). This corresponds to about two thirds of the present annual increment of atmospheric carbon dioxide which amounts to about eight thousand million tonnes (2.2 thousand million tonnes of carbon).

It would, of course, take several decades until the newly planted forests reach such a level of productivity. On the other hand, the sink would continue to exist for about half a century, i.e. as long as the new forests are growing, and then it would only gradually fade out as they reach maturity.

If in addition it were possible to put an end to the rapid destruction of existing forests and, as a first step, to compel the timber corporations and charcoal producers to plant in their working areas at least as many new trees as they cut, a carbon dioxide source now about as large as that from the burning of fossil fuel could at least be considerably reduced. Summing the atmospheric effects of reforestation and the preservation of existing forests, there can be no doubt that a reasonable forest policy, though not a panacea, could *significantly* contribute to slow down the increase in the atmospheric carbon dioxide burden.

There is no question that a worldwide reforestation campaign would be technically feasible, though it would certainly take some time to gather the necessary experiences and find a proper way of blending modern science and technology with traditional local knowledge. Moreover, such a campaign would provide many jobs for the rural population in more or less backward countries suffering from permanent structural unemployment. To foster such schemes would be one of the most sensible forms of development aid and, by the way, *a perfectly moral way of helping oneself by helping others.*

There would be no need for comprehensive international agreements to make a start and get some pilot projects going. A country providing development aid could earmark a certain amount of its funds for forestry schemes and let it be known that recipient countries willing to co-operate in such schemes would get preferential treatment. Reforesting about three million square kilometres of now denuded land would of course cost a lot of money. But in the long run, the scheme would pay for itself, *even disregarding the effects on the carbon dioxide problem.* It would be difficult to provide exact cost/benefit calculations, for the main benefits would be the prevention of losses by water and wind erosion, flooding and drought, landslides and avalanches, and the prevention of silting-up of reservoirs and irrigation channels.[28] The value of such losses not incurred can only be estimated in a very crude way. But every person with knowledge of the facts would agree that these benefits are very real indeed and not just imaginary.

Reforestation would use up some land that is now under, albeit very precariously, cultivation. But done in a proper way, concentrating on mountain slopes, hilltops and the creation of windbreak belts in the plains, it would by its beneficial effects *increase* overall agricultural production.

It will be even more difficult to *prove* the benefits of mitigating the carbon dioxide problem. For while there are many people who know from their own dire experiences the terrible local consequences of erosion etc. which can be prevented by reforestation, nobody has as yet experienced the consequences of increased carbon dioxide concentrations. And scientists, while on the whole agreeing that this might constitute a real and perhaps very grave danger, cannot predict except in a very vague and general way what these dangers might be.

Thus the main problem of large-scale reforestation might be the *political* feasibility. Will it be possible to convince the politicians and the tax-payers of the rich countries to invest considerable sums of money in such projects, the benefits of which will only accrue in several decades? Will it be possible to convince them that scientists are right when they say that these benefits are worldwide inasmuch as they can prevent possible climatic catastrophes which might hit our children's grandchildren's generations all over the world?

But this question of the political feasibility is not a special problem of reforestation. It is bound to affect every energy programme except perhaps the simple 'wait and see' policy. To prevent a carbon dioxide build-up to what are probably dangerous levels, it would not be sufficient to just subsidise and speed up the development and deployment of fast breeders, the methanol scheme, solar hydrogen production or some other alternative and leave the rest to the free play of the market forces. 'To make any dent at all in the carbon dioxide problem will require a sustained coherent policy enforced literally over a period of centuries, and on a worldwide basis,' writes H. Brooks in his paper presented at the Dahlem conference.[29] And at the IIASA workshop G. Zimmermeyer struck the same note when he asked: 'Where is the institution that could impose equal restrictions on carbon dioxide emissions, perhaps by rationing the use of fuel or by forbidding certain technical processes; restrictions that would be valid for *all* countries, especially those still in development, where the availability of fossil fuel energy is vital for man and his health?'[30]

Seen in this light a worldwide reforestation campaign, with its *multiple* benefits, might still be easier to implement than any other strategy to mitigate the carbon dioxide problem.

After us the deluge?

'What has posterity done for me, that I should do something for posterity?' said George Bernard Shaw. The great satirist had certainly not raised this question with regard to environmental considerations or afforestation. However it fits so well the problems dealt with in this book.

The consequences of the increase in carbon dioxide or the depletion in ozone will be noticeable in several decades at the earliest. If for the time being no action is taken there will be no environmental risk for the immediate future. Thus is it really at all worthwhile to deal with these problems – 'for posterity'? Are there not already enough urgent problems that embitter our lives? To solve such problems only limited funds are available and the capacity of public interest is limited too. Thus is it not better merely to focus just on the most urgent problems and postpone all the others?

But the things embittering our lives now have not always been so urgent. Today we have to deal with these problems because our forefathers postponed them some decades ago. Should we do the same? The difference is however that we shall leave even more difficult problems for posterity, since the technological potential and thus the tendency to destroy nature becomes larger, and hence solving environmental problems becomes more difficult, the more people live on Earth.

The search for a reasonable compromise between the interests of present and future generations is a difficult task. The reason is not only that politicians, who after all make the decisions and provide the funds, can hardly be interested in problems that are of no importance for the next election, or even the one after that. We all have legitimate selfish interests that can be in conflict with those of future generations. Are we really willing to make a sacrifice for posterity or just merely to relinquish some comfort?

The solution to such environmental problems is made even more difficult by the fact that often decisions have to be made already at times when still no reliable basis for the decision exists. 'The need for some measure of action usually precedes the accumulation of definite evidence as to the scope and magnitude of a postulated threat.'[31]

The presumption that such a risk may exist at all is usually based upon considerations that are certainly recognised by the majority of the scientists as being plausible, but that have still not been proved. For the public and the politicians such considerations are sometimes difficult to grasp. Often the assessments are based, as has been repeatedly pointed out in this book, on equations with many unknowns, supplemented by certain assumptions. Even more uncertain than the prediction of likely changes in physical or chemical properties, like the carbon dioxide concentration in the atmosphere for instance, is the judgement of the medical, ecological and climatological consequences, and more uncertain still is the assessment of how and to what extent climatological changes will effect agriculture and human life in general.

Under these circumstances politicians may certainly be tempted to postpone the decisions until the experts are in better agreement. This holds true in particular when the necessary action would cost much money or require a considerable disruption to the economy. But even the decision to do nothing for the time being is a decision that possibly might have serious consequences some time next century.

It is especially difficult to properly classify a newly recognised – or at least suspected – danger among the entirety of environmental problems. In times of sensational headlines, a moderately balanced, objective presentation might often fail to attract public attention. But until a certain threshold of awareness of the public and the politicians is reached, there is no chance at all that any action will be taken beyond a mere researching of the problem. On the other hand, however, unduly exaggerating a (possible) risk may lead to unreasonable and hasty over-reactions that might even make the solution more difficult, or solve one problem by introducing a bigger one. This risk is even greater, because there are always lobbies that eagerly play up certain questions (or sweep them under the carpet, when it suits them) in order to further their own interests.

Many scientists are convinced that they are capable of finding the right and objective solutions to the long-term environmental problems, if they, and hence the scientific community at large, were charged with the responsibility of this. But this, says Brooks, is a mere technocratic prejudice. There can be no rational 'objectively correct' solutions for balancing different conflicting interests of the present and of the future one against the other. This is not a search for the 'truth' but a process of negotiation in order to achieve a compromise that is as sound as possible, and that will not have to be revised again and again in the years to come. 'Many argue', he continues, 'that any human activity which produces effects large when compared with the corresponding natural cycles should be viewed with great suspicion. While this may be true for any single environmental threat, if we try to apply this principle across the board, we may find ourselves with limitations on economic activity which cannot reasonably be tolerated. There is just not that much elbow room for superconservatism with respect to risks within the workings of the economic system, especially when one takes into account the poor.'[32]

Seen that way the solution to environmental problems is far more than the search for a compromise between present and future interests. In the interests of future generations that might number some eight, ten or even twelve thousand million people, we have to provide the preconditions right now for a *manifold* increase in future production of food, raw materials and energy, and this will undoubtably have a severe impact on our environment.

On the one hand, the dire needs of more and more people in future generations threaten; on the other, the risk of completely destroying our environment. If we are lucky a narrow path may perhaps be found which leads between these two evils.

Tables

Table 1. *Composition of the Earth's atmosphere.*

Gas	Volume (%)	Mass (%)	Mass (thousand million tonnes)
Nitrogen (N_2)	78.09	75.51	3 865 000
Oxygen (O_2)	20.95	23.15	1 184 000
Argon (A)	0.93	1.28	65 000
Carbon dioxide* (CO_2)	0.03	0.046	2 350
(CO_2 = 1975)*	0.0326	0.051	2 600

Trace gases	Volume (millionths)	Mass (millionths)	Mass (thousand million tonnes)
Neon (Ne)	18	12.5	64
Helium (He)	5.2	0.72	3.7
Methane (CH_4)	1.5	0.94	4.3
Krypton (Kr)	1.0	2.9	15
Laughing gas (N_2O)**	0.5	0.8	4
Hydrogen (H_2)	0.5	0.035	0.18
Ozone (O_3)	0.4	0.7	3.3
Xenon (Xe)	0.08	0.36	1.8
In addition varying amounts (up to 2%) of water vapour			

*With seasonal variations of several per cent.

**According to recent measurements (see page 119), the content of laughing gas in the atmosphere amounts to about 0.25 to 0.3 millionths by volume corresponding to a total mass of 1.5 to 2 thousand million tonnes.

From G. E. Hutchinson in G. P. Kuiper (ed.) *The Earth as a Planet*, Chicago 1954, p. 389.

Table 2 *Net primary production and biomass of various ecosystems, (figures in amounts of carbon*).*

	Area (million km²)	Net primary production: Range (g/m² per year)	Net primary production: Average (g/m² per year)	Net primary production: Total (thousand million tonnes)	Biomass: Range (kg/m²)	Biomass: Average (kg/m²)	Biomass: Total (thousand million tonnes)
Tropical rain forest**	17	450 –2250	900	16.8	3–36	20	340
Tropical seasonal forest	7.5		720	5.4		16	120
Temperate evergreen forest	5	270 –1350	585	2.9	3–90	16	80
Temperate deciduous forest	7		540	3.8		13.5	95
Boreal forest	12	180 –900	360	4.3	3–18	9	108
Total forest	*48.5*		*684*	*33.2*		*15.3*	*743*
Woodland and shrub	8.5	90 –540	320	2.7	1–9	2.7	22
Savannah	15	90 –900	410	6.1	0.1–6.7	1.8	27
Temperate grassland	9	65 –675	265	2.4	0.1–2.3	0.7	6.3
Tundra and alpine grassland	8	4.5–180	65	0.5	0.05–1.4	0.3	2.4
Desert scrub	18	4.5–110	40	0.7	0.05–1.8	0.33	5.9
Desert (sand, rock, ice)***	24	0 –5	1.4	0.03	0–0.1	0.01	0.2
Cultivated land	14	45 –1800	290	4.1	0.2–5.4	0.5	7
Swamp and marsh	2	360 –1800	1350	2.7	1.5–22.5	6.8	13.6
Lake and river	2	45 –675	200	0.4	0–0.05	0.01	0.02
Total land	*149*		*355*	*52.8*		*5.55*	*827.5*

Ocean	332	1– 360	56.5	18.7	0–0.002	0.0014	0.45
Upwelling zones	0.4		225	0.1		0.01	0.004
		90– 270			0–0.02		
Continental shelf	26.6		162	4.3		0.00 5	0.12
Reef	0.6		1000	0.6		0.9	0.54
		225–1800			0.02–1.8		
Estuary	1.4		810	1.1		0.45	0.63
Total oceans	*361*		*69*	*24.8*		*0.005*	*1.74*
Total Earth	510		152	77.5		1.63	829

From Whittaker and Likens. Figures correspond to the year 1950. Whittaker and Likens have repeatedly revised their table. See R. H. Whittaker *Communities and Ecosystem*, New York 1970; R. H. Whittaker and G. E. Likens in: G. M. Woodwell and E. V. Pecan (eds.), *Carbon and the Biosphere*, Springfield, Va., 1973; R. H. Whittaker and G. E. Likens in: H. Lieth and R. H. Whittaker (Eds.), *Primary Productivity of the Biosphere*, New York 1975. Similar figures are also found in H. Lieth, *Umschau 74*, 171, 1974.

*The figures correspond to the carbon content of the organic substance, which accounts for about 45 per cent of the dry mass.
**According to the World Wildlife Fund, only 9.5 million square kilometres of rain forest were left in 1975 (see page 79).
***According to Lieth (op. cit.), 15.5 million of the 24 million square kilometres of desert is ice desert where practically no primary production takes place.

Notes

Is our atmosphere endangered?

1 In the near-surface layers of the atmosphere the water vapour content varies between 0 and 2 per cent. In the stratosphere only trace amounts of water vapour (some tenthousandths of a per cent) exist.
2 R. M. Garrels and A. Lerman in: W. Stumm (ed.), *Global Chemical Cycles and their Alterations by Man*, Dahlem Conference, Berlin 1977, p. 30.
3 C. C. Delwiche, *Scientific American*, September 1970, p. 137.
4 P. Cloud and A. Gibor, *Scientific American*, September 1970, p. 111.
5 J. Williams (ed.), *Carbon Dioxide, Climate and Society, proceedings of an IIASA workshop*, Pergamon Press, Oxford, 1978.

1 Oxygen and carbon dioxide – wherein lies the danger?

1 P. Böger, *Naturw. Rdsch.* **28**, 429, 1975.
2 B. Bolin, *Scientific American*, September 1970, p. 125.
3 See, for instance, C. A. Ekdahl and C. D. Keeling in: G. M. Woodwell and E. V. Pecan (eds.), *Carbon and the Biosphere*, USAEC conference-720510 (24th Brookhaven symposium in biology), Springfield, Va. 1973; see also Bolin.
4 See, for instance, R. M. Garrels, A. Lerman and F. T. Mackenzie, *American Scientist* **64**, 306, in particular 308, 1976.
5 According to R. H. Whittaker, *Communities and Ecosystem*, New York 1970, p. 83. See also table on pages 172–3 of this book.
6 The importance of humus with regard to the carbon cycle is discussed in part III, page 80.
7 See J. Weck, *Die Wälder der Erde*, Göttingen 1957, p. 99. See also *New Scientist* **72**, 4, 1976, and **75**, 582, 1977, (report about UNCOD conference).
8 S. Rehm, *Umschau* **73**, 44, 1973.
9 A. Gomez-Pompa *et al.*, *Science* **177**, 762, 1972.
10 R. H. Whittaker and G. E. Likens in: Woodwell and Pecan (eds.) p. 295.
11 E. F. Brünig, *Umschau* **74**, 405, 1974.
12 J. A. S. Adams *et al.*, *Science* **196**, 54, 1977.
13 H. Grabert, *Umschau* **74**, 246, 1974.

14 Rehm.
15 K. Openshaw, *New Scientist* **61**, 271, 1974.
16 A. Agarwal, *New Scientist* **67**, 386, 1975.
17 H. Ern, *Umschau* **73**, 85, 1973.
18 Rehm.
19 Gomez-Pompa *et al.*
20 R. Allen, *New Scientist* **66**, 178, 1975.
21 E. F. Brünig, *Ambio* **6**, 190, 1977 (slightly corrected).
22 See, for instance, M. E. Adzhiev, *Priroda* **8**, 1975. Excerpts of this article were translated and published in *New Scientist* **69**, 382, 1976. The situation in Canada is described in *New Scientist* **73**, 24, 1977.
23 See, for instance, G. M. Woodwell and R. A. Houghton in: W. Stumm (ed.), p. 61; also G. M. Woodwell, *Scientific American*, January 1978, p. 34.
24 Openshaw.
25 J. Tinker, *New Scientist* **63**, 819, 1974.
26 Agarwal.
27 Openshaw.
28 Agarwal.
29 N. Myers, *New Scientist* **80**, 951, 1978.
30 K. Heine, *Umschau* **76**, 202, 1976.
31 W. Sandermann, *Umschau* **73**, 51, 1973. For a discussion on the environmental impact of erosion in general terms, see, for instance, E. P. Eckholm, *Losing Ground*, Norton, New York, 1976.
32 S. Judson, *Science* **160**, 1444, 1968.
33 S. Judson, *American Scientist* **56**, 356, 1968.
34 S. Judson, *American Scientist*; B. Gregor, *Nature* **228**, 273, 1970.
35 R. M. Garrels and A. Lerman in: Stumm (ed.), p. 26.
36 See, for instance, R. H. Whittaker and G. E. Likens in: Woodwell and Pecan (eds.), p. 281.
37 Ibid.
38 Garrels *et al.*, *American Scientist.*
39 G. Breuer, *Die Herausforderung; Energie für die Zukunft – Gefahren und Möglichkeiten.* C. Bertelsmann, Munich 1975, p. 29.
40 See, for instance, C. Cole LaMont, *Umschau* **69**, 557, 1969; *Naturw. Rdsch.* **27**, 74, 1972; Th. Keller, *Schweiz. Ztschr. f. Forstwesen*, **124**, 465, 1973.
41 See, for instance, W. S. Broecker, *Science* **168**, 1537, 1970. According to estimates of other authors, the reserves of fossil fuel are smaller. Thus two per cent of atmospheric oxygen would be sufficient to burn them.
42 See, for instance, F. D. Sisler in: S. F. Singer (ed.), *The Changing Global Environment*, Reidel, Dordrecht, Boston 1975, p. 59.
43 C. Cole LaMont, *Bioscience* **18**, 679, 1968.

44 Keller.
45 J. H. Ryther, *Nature* **227**, 374, 1970; Broecker.
46 H. D. Holland, *Ocean Waters, Nutrients and Atmospheric Oxygen* proceedings of the symposium on hydrogeochemistry and biochemistry, Washington, 1973, p. 68, in particular p. 69. See also M. Schidlowski, R. Eichmann and C. E. Junge, *Precambrian Research* **2**, 1975, in particular p. 55. According to model calculations made by Garrels *et al.* (*American Scientist*) about twelve million years would be required to use up all the oxygen present in the atmosphere through geological processes.
47 Garrels *et al.*, *American Scientist.*

2 The history of the atmosphere

1 See, for instance, H. E. Suess, *Journal of Geology* **57**, 600, 1949; see also H. Brown in: G. P. Kuiper (ed.) *The Atmospheres of the Earth and the Planets*, Chicago 1949.
2 See M. Schidlowski in: B. F. Windley (ed.), *The Early History of the Earth*, Wiley, London 1976, pp. 525–35; see also J. C. G. Walker, ibid. pp. 537–46; M. Schidlowski, *Origin of Life*, proceedings of 2nd ISSOL conference, Center. Acad. Publ. Japan, Tokyo 1978, pp. 3–20.
3 According to H. D. Holland in: P. J. Brancazio and A. G. W. Cameron (eds.), *The Origin and Evolution of Atmospheres and Oceans*, Wiley, New York 1964, p. 93.
4 Holland, p. 96.
5 H. C. Urey, *The Planets: Their Origin and Development*, Yale University Press, New Haven 1953, p. 148.
6 P. Ramdohr, *Abhdlg. Deutsche Akad. d. Wiss. Berlin* 1958, *Kl. Chemie, Geologie und Biologie* **3**, 1; see also M. G. Rutten, *The Geological Aspects of the Origin of Life on Earth*, Elsevier, Amsterdam 1962, p. 96.
7 Rutten, pp. 97 and 109; P. Cloud and A. Gibor, *Scientific American*, September 1970, p. 111.
8 See L. V. Berkner and L. C. Marshall in: Brancazio and Cameron (eds.), p. 112.
9 R. T. Brinkmann, *Journal of Geophysical Research* **74**, 5355, 1969.
10 See M. Schidlowski in: Windley (ed.), p. 525.
11 A. I. Opàrin, *The Origin of Life*, Macmillan, New York 1938. Oparin first started to deal with this subject in 1924. See also J. B. S. Haldane, *The Origin of Life*, Harper, New York 1933.
12 S. L. Miller, *Science* **117**, 528, 1953.
13 See review article by K. Dose, *Unschau* **67**, 683, 1967; S. W. Fox, *The Origin of Prebiologic Systems*, Academic Press, New York, 1965.
14 See J. Monod, *Chance and Necessity.*
15 M. Eigen, Foreword to Monod, German edition p. xv.

16 Monod.
17 R. W. Kaplan, *Umschau* **72**, 456, 1972; see also *Naturw. Rdsch.* **30**, 197, 1977.
18 See N. W. Pirie in: A. I. Oparin (ed.), *The Origin of Life on Earth*, I, Pergamon, London 1959, p. 76; see also Rutten, p. 129.
19 S. Hill, *Journal of General Microbiology* **67**, 77, 1971.
20 Rutten, p. 76; see also J. W. Schopf *et al.*, *Journal of Palaeontology* **45**, 477, 1971; G. Bond *et al.*, *Nature* **244**, 275, 1973.
21 J. W. Schopf in: Windley (ed.), p. 590; see also D. M. McKirdy, *Precambrian Research* **1**, 75–137, 1974.
22 See S. A. Tyler and E. S. Barghoorn, *Science* **119**, 606, 1954; also *Science* **147**, 563, 1964; see also E. S. Barghoorn, J. W. Schopf *et al.*, *Science* **148**, 77, 1964; also *Science* **149**, 1365, 1964.
23 See J. Hoefs and M. Schidlowski, *Science* **155**, 1096, 1967; A. Prashnowsky and M. Schidlowski, *Nature* **216**, 560, 1967.
24 E. S. Barghoorn and J. W. Schopf, *Science* **152**, 758, 1966; *Science* **156**, 508, 1967; see also M. D. Muir and P. R. Grant in: Windley (ed.), p. 595.
25 A. H. Knoll and E. S. Barghoorn, *Science* **198**, 396, 1977.
26 See, for instance, M. Schidlowski, *Umschau* **68**, 566, 1968.
27 An extensive report about these investigations is given by M. Schidlowski *et al.*, *Precambrian Research* **2**, 1, 1975.
28 P. Cloud, *Science* **160**, 729, 1968; *Paleobiology* **2**, 351, 1976; see also Cloud and Gibor.
29 S. Moorbath *et al.*, *Nature* **245**, 138, 1973.
30 M. Schidlowski *et al.*, *Geochimica Cosmochimica Acta* **43**, 189–99, 1979.
31 *MPG Presse information*, 20 June 1979.
32 See Berkner and Marshall, p. 115; see also A. G. Fischer, *Proceedings of the National Academy of Sciences of the USA* **53**, June 1965, symposium on the evolution of the atmosphere, p. 1208.
33 For the energy transfer adenosintriphosphate (ATP) plays the role of 'small change'. Thus the number of ATP molecules formed is the crucial factor for an energy producing reaction. Through fermentation processes, one molecule of glucose produces two molecules of ATP, while through respiration thirty-eight ATP molecules are formed. With respect to calories, respiration yields 'only' fourteen times as much energy as compared to fermentation. This energy, however, is more effectively used in the respiration process. Thus, in the end, nineteen times as many ATP molecules are produced.
34 J. W. Schopf and D. Oehler, *Science* **193**, 47, 1976.
35 R. H. Whittaker, *Communities and Ecosystems*, New York 1970, p. 142; J. C. G. Walker, *American Journal of Science* **274**, 193, 1974; Schidlowski *et al.*, *Precambrian Research*; Garrels *et al.*
36 Walker, p. 202.

37 See Schidlowski *et al.*, *Precambrian Research*, pp. 45 and 54.
38 See Garrels *et al.*
39 Ibid.

3 Carbon dioxide increase – how great is the risk?

1 See, C. A. Ekdahl, C. D. Keeling, op. cit. (see part I, ref. 3); B. Bolin and W. Bischof, *Tellus* **22**, 431 1970; W. Bischof, Report AC-36, Department of Meteorology, University of Stockholm, 1976. See also G. M. Woodwell, *Scientific American*, January 1978, p. 34.
2 G. S. Callendar, *Tellus* **10**, 243, 1958.
3 In 1954 already G. E. Hutchinson pointed out that the increase of atmospheric carbon dioxide concentration is partly due to forest clearing and humus destruction. However, this hint has obviously not been taken note of. (See G. E. Hutchinson in: G. P. Kuiper (ed.), *The Earth as a Planet*, Chicago 1954, p. 389.)
4 C. D. Keeling, *Tellus* **25**, 174, 1973.
5 See M. Chatterjee (India), 8th world energy conference, Bukarest 1971; K. Openshaw, *New Scientist* **61**, 271, 1974.
6 Based on extrapolations of figures from various countries C. S. Wong assessed that in temperate and boreal zones some 1.7 thousand million tonnes of carbon dioxide (containing 470 million tonnes of carbon) are produced by forest fires every year. This would imply that about four per cent of the organic substance produced by these forests are oxidised that way. In the tropics, in particular in rain forests, he assumes a rarer incidence of forest fires. Due to slash-and-burn agriculture (shifting cultivation) however, some 13.5 to 19 thousand million tonnes of carbon dioxide (containing 3.7 to 5.2 thousand million tonnes of carbon) are produced there every year. Due to the growth of the population, increasingly larger areas are being involved in this shifting agriculture. (*Science* **200**, 197, 1978.)
7 See H. E. Suess, *Science* **122**, 414, 1955).
8 Bolin and Bischof.
9 L. Machta in: D. Dyrssen and D. Jagner (eds.) *The Changing Chemistry of the Oceans*, Wiley, New York 1972, p. 121.
10 R. Bacastow and C. D. Keeling in: G. M. Woodwell and E. V. Pecan (eds.), *Carbon and the Biosphere*, Springfield, Va. 1973, p. 86.
11 G. M. Woodwell, R. H. Whittaker, W. A. Reiners, G. E. Likens, C. C. Delwiche and D. B. Botkin, *Science* **199**, 141, 1978.
12 See also G. M. Woodwell and R. A. Houghton in: W. Stumm (ed.), *Global Chemical Cycles and their Alterations by Man*, Berlin 1977, p. 69.
13 See Woodwell and Pecan (eds.).
14 ERDA workshop on *Environmental Effects of Carbon Dioxide from*

Fossil Fuel Combustion, Miami, Florida, March 1977. (Proceedings, W. P. Elliot and L. Machta (eds.), US Department of Energy, Washington, 1977.)

ICSU-Scope conference on *The Global Carbon Cycle*, Ratzeburg, German Federal Republic, April 1977. (Proceedings, B. Bolin, E. T. Degens, S. Kempe and P. Ketner (eds.), Wiley, Chichester, New York, 1978.)

IIASA workshop on *Carbon Dioxide, Climate, and Society*, Baden, Austria, February 1978 (Proceedings, J. Williams (ed.), Pergamon Press, Oxford 1978.)

Conference on *Man's Impact on Climate*, Berlin, June 1978. (Proceedings, W. Bach (ed.) Elsevier, Amsterdam 1979.)

15 See B. Bolin in: Williams (ed.), p. 41; and A Björkström, ibid., p. 97.

16 Woodwell, *Scientific American.*

17 Quoted by Woodwell *et al.*, *Science.*

18 W. A. Reiners in: Woodwell and Pecan (eds.), p. 303.

19 W. H. Schlesinger, *Annual Review of Ecological Systems*, 8, 51 (1977); H. L. Bohn, *Soil Science Soc. Am. J.* 40, 468 (1976).

20 See also S. Kempe in: Bolin *et al.* (eds.), Ratzeburg conference.

21 M. Stuiver, *Science* **199**, 253 (1978).

22 H. D. Freyer in: Williams (ed.), p. 69.

23 B. Bolin in: Williams (ed.), p. 41.

24 H. Brown and F. Escombe, *Proceedings of the Royal Society, London*, Ser. B, **76**, 118, 1905.

25 G. M. Woodwell and R. A. Houghton in: Stumm (ed.); Woodwell and Whittaker *et al.*, *Science.*

26 Ibid.

27 Wong (see ref. 6).

28 Woodwell *et al.*

29 E. T. Alexandersson, *Nature* **262**, 653, 1976; *Geology* **6**, 324, 1978.

30 K. E. Zimen in: Williams (ed.), p. 89.

31 P. Abelson and T. Malone (eds.), *Energy and Climate: Outer Limits to Growth?*, National Academy of Sciences of the USA, Washington, 1977.

32 G. M. Woodwell in: Williams (ed.), pp. 65–6.

33 R. Bacastow and C. D. Keeling in: Woodwell and Pecan (eds.); R. E. Hamilton in: Stumm (ed.), p. 155.

34 K. E. Zimen and F. K. Altenhein, *Ztschr. f. Naturfschg.* **28a**, 1747, 1973; K. E. Zimen *et al.*, ibid. **32a**, 1544, (1977).

35 See K. E. Zimen in: Williams (ed.), p. 89.

36 B. Bolin, *Science* **196**, 613, 1977.

37 B. Bolin in: Williams (ed.), p. 41.

38 See A. Nir *et al.* (group report) in: Stumm (ed.), p. 316.

39 M. I. Budyko and K. Y. Vinnikow in: Stumm (ed.), p. 189.

40 L. Machta in: W. N. Hess (ed.), *Weather and Climate Modification*, Wiley, New York, 1974, p. 700; H. Flohn in: Stumm (ed.), p. 211.
41 H. Flohn in: Williams (ed.), p. 227; V. Ramanathan, *Science* **190**, 150, 1975; W. D. Wang *et al.*, *Science* **194**, 685, 1976; *WMO* 135, 6 January 1976.
42 Budyko and Vinnikow.
43 H. Flohn in: Stumm (ed.), p. 207; J. J. Morgan *et al.* (group report), ibid. p. 298.
44 L. Machta in: Woodwell and Pecan (eds.), p. 21.
45 W. W. Kellogg, IIASA workshop preprints, see also W. W. Kellogg *Bulletin of the Atomic Scientists* (in print).
46 See Williams (ed.), report of working group II, p. 307.
47 Kellogg.
48 See U. Hampicke, *Umschau* **77**, 599, 1977.
49 J. O. Fletcher, *Bulletin of the Atomic Scientists*, December 1970, p. 39.
50 K. Aagaard and L. K. Coachman, *Eos* **56**, 484, 1975.
51 Kellogg; see also G. Breuer, *Weather modification: prospects and problems*, Cambridge University Press 1980, p. 106.
52 A. T. Wilson, *Nature* **201**, 147, 1964; T. Hughes, *Journal of Geophysical Research* **78**, 7884, 1973.

4 Nitrogen and ozone – unexpected relationships

1 See E. Broda, *Journal of Molecular Evolution* **7**, 87, 1975.
2 About nitrogenase, see R. C. Burns and R. W. F. Hardy, *Nitrogen Fixation in Bacteria and Higher Plants*, Springer, Heidelberg 1975, p. 61; R. W. F. Hardy and U. D. Havelka, *Science* **188**, 633, 1975.
3 See also E. Broda, *Origins of Life* **8**, 87, 1977.
4 R. A. Dixon and J. R. Postgate, *Nature* **237**, 102, 1972.
5 Burns and Hardy, p. 18.
6 R. H. Whittaker, *Communities and Ecosystems*, New York 1970, p. 111.
7 Burns and Hardy, p. 28.
8 See J. F. W. v. Bülow and J. Döbereiner, *Proceedings of the National Academy of Sciences of the USA* **72**, 2389, 1975.
9 Hardy and Havelka.
10 J. Hahn and C. Junge, *Ztschr. f. Naturfschg.* **32a**, 190, 1977, based on F. B. Cady and W. v. Batholomew, *Soil Sciences Society Proceedings* **25**, 477, 1961.
11 Whittaker, p. 120.
12 C. C. Delwiche, *Scientific American*, September 1970, p. 111.
13 See C. C. Delwiche and G. E. Likens in: W. Stumm (ed.), *Global Chemical Cycles and their Alterations by Man*, Berlin 1977, p. 78; see also T. W. Ashenden and T. A. Mansfield, *Nature* **273**, 143, 1978.
14 See Delwiche; Hardy and Havelka, working for the research and development laboratory of the American chemical corporation Du

Pont de Nemours, even consider 200 million tonnes until the year 2000.

15 See C. C. Delwiche, *Ambio* **6**, 106, 1977, in particular p. 110; see also Burns and Hardy, p. 43.

16 See H. J. Simpson *et al.* (group report) in: Stumm (ed.), p. 253, in particular p. 264.

17 See Hahn and Junge; J. Hahn in: W. Bach (ed.), *Man's Impact on Climate*, Berlin 1978. From these papers also many other figures for this section were taken.

18 E. C. Zipf, Annual Meeting, *American Geophys. Union*, 1975.

19 See C. Junge, *Quarterly Journal of the Royal Meteorological Society* **98**, 711, 1972; H. S. Johnston and G. Selwyn, *Geophysical Research Letters* **2**, 549, 1975; P. J. Crutzen, *Geophysical Research Letters* **3**, 1969, 1976.

20 See H. J. Simpson *et al.* (group report) in: Stumm (ed.), p. 263.

21 See Hahn and Junge.

22 See F. Timmermann, *Umschau* **77**, 314, 1977.

23 See *Halocarbons – Environmental Effects of Chlorofluoromethane Release*, National Academy of Sciences of the USA, Washington 1976, p. 316.

24 See, for example, P. Fabian, *Nature* **272**, 306, 1978.

25 J. Hahn in: Bach (ed.).

26 See N. Valery, *New Scientist* **68**, 19, 1975; Fabian.

27 M. Sherwood, *New Scientist* **68**, 15, 1975.

28 M. J. Molina and F. S. Rowland, *Nature* **249**, 810, 1974; see also S. Rowland, *New Scientist* **64**, 717, 1974.

29 *Halocarbons*, p. 5/4.

30 Fabian.

31 *Halocarbons*, pp. 5/19 and 5/13.

32 T. Stoel, *New Scientist* **81**, 166, 1979.

33 Ibid.; for a comprehensive review of the spray-can issue see L. Dotto and H. Schiff, *Ozone War*, Doubleday, New York 1978.

34 See A. C. Giese, *Living with our Sun's Ultraviolet Rays*, Plenum Press, New York 1976, p. 173.

35 Quoted from Rowland, 1974 and *New Scientist* **68**, 8, 1975.

36 Rowland, 1975; see also Giese, p. 175.

37 H. Brooks in: Stumm (ed.), p. 246.

38 See Rowland, 1975, with reference to a study of the American Academy of Sciences.

5 Between the devil and the deep blue sea

1 K. E. Jay, *Calder Hall*, British Atomic Energy Depertment 1956; see also K. E. Jay, *Britain's Atomic Factories: The Story of Atomic Energy Production in Britain*, HMSO 1954.

2 See E. Broda, *Naturw. Rdsch.* **28**, 233, 1975.

3 *24th Pugwash Conference on Science and World Affairs*, Baden, Austria, August/September 1974. The Pugwash conferences are non-official discussions of leading scientists from West and East, and the Third World, dealing with problems of nuclear weapons and disarmament in particular. They contributed substantially in preparing the nuclear weapon test stop treaty. Pugwash is a Canadian health spa where the first of these conferences was held in 1957.
4 Ford Foundation/MITRE Corporation, *Nuclear Power: Issues and Choices*, Ballinger, Cambridge, Mass, 1977; see also F. Barnaby, director of the Stockholm International Peace Research Institute (SIPRI), *New Scientist* **75**, 168, 1977.
5 See the often-quoted 'Flowers Report', *Nuclear Power and the Environment*, HMSO, Comm. Paper Nr. 6618, London 1976.
6 H. Flohn, *Umschau* **77**, 569, 1977.
7 U. Hampicke, *Umschau* **77**, 605, 1977.
8 See, for instance, Energy Policy Project of the Ford Foundation, *A Time to Choose*, Ballinger, Cambridge, Mass. 1976, p. 136.
9 A. M. Weinberg and R. P. Hammond, *Bulletin of the Atomic Scientists*, March 1972, p. 6.
10 A. M. Weinberg and R. M. Rotty in W. Stumm (ed.), op. cit., p. 225, The energy forecast is based on a paper of F. Niehaus, *A Non-linear eight Level Tandem Model to calculate the Future CO_2 burden ...*, RM-36-35, IIASA, 1976.
11 W. Häfele in: J. Williams (ed.), *Carbon Dioxide, Climate and Society*, Oxford 1978, p. 21.
12 G. Zimmermeyer in: Williams (ed.), p. 275.
13 J. R. Ravetz, *Scientific Knowledge and its Social Problems*, Clarendon, Oxford 1971, p. 348.
14 See *Bild der Wissenschaft* **3**, 64, 1977.
15 A. M. Weinberg and R. M. Rotty in: W. Stumm (ed.), *Global Chemical Cycles and their Alterations by Man*, Berlin 1977, p. 225.
16 See E. Broda, *Bulletin of the Atomic Scientists* 3, 49, 1976.
17 US Office of Technology Assessment, *Application of Solar Technology of Today's Energy Needs*, Washington 1978.
18 Häfele.
19 Report group III in: Williams (ed.), p. 315.
20 Häfele.
21 Häfele; C. Mustacchi *et al.* in: Williams (ed.), p. 283; C. Marchetti, *Climatic Change* **1**, 59, 1977.
22 A. Nir *et al.* (group report) in: Stumm (ed.), pp. 320, 321.
23 B. Schefold in: Stumm (ed.), p. 327.
24 J. Giono, lecture before the English Society, 'The men of the trees' (French original title 'Un caractère'; in German: 'Der Mann mit den Bäumen'). German translation by W. Tappolet, Zürich 1972, in particular pp. 18, 27, 30.

25 Ibid., cited in the epilogue of the translator.
26 Ibid. p. 13.
27 For a report on a seemingly very successful reforestation in South Korea, see E. Eckholm, *New Scientist* **82**, 764, 1979.
28 According to a report presented at the 8th World Forestry Conference in Jakarta the life of seventy-one silting-up reservoirs in the Poza Honda River basin in Ecuador could be extended by an average of twenty-five years by watershed conservation measures costing one million dollars and producing benefits worth two million. N. Myers, *New Scientist* **80**, 951, 1978.
29 H. Brooks in: Stumm (ed.), p. 249.
30 Zimmermeyer.
31 Brooks, p. 245.
32 Brooks, p. 247.

Glossary of technical terms

Anaerobes: Unicellular organisms that do not need oxygen and that meet their energy requirements through fermentation. Facultative anaerobes are capable of respiration, but can switch to other methods of energy production if oxygen is deficient. Obligatory anaerobes cannot respire at all, and most of them can thrive only in the absence of oxygen.

Autotrophs: Organisms that take up carbon as carbon dioxide, converting it to higher organic compounds, in particular the green plants. The opposite: heterotrophs.

Biomass: The total living substance (plants, animals, microorganisms) that exist in a certain environment or on the whole globe. Some authors inaccurately also include the decaying organic substance (humus for instance).

Carbohydrates: Compounds of carbon, hydrogen and oxygen, the latter two being in the ratio 2:1. These are products of photosynthesis storing solar energy as chemical energy: glucose, starch, cellulose.

Catalyst: Substance that, by its mere presence, makes certain chemical reactions proceed faster, without becoming part of the reaction's end product.

Denitrifying bacteria: See nitrate respirers.

Ecology: Science of life and relationships of different unicellular and multicellular organisms existing together in a common environment (ecosystem), a forest for example.

Erosion: Removal of topsoil by water, ice or wind.

Eukaryotes: Organisms that possess a nucleus in their cells. The opposite: prokaryotes. All multicellular organisms are eukaryotes.

Fossil fuels: Fuels deposited in the Earth's crust, originating from plants or animals of past geological ages and storing the solar energy of that time as chemical energy (coal, oil, natural gas).

Half-life (period): The time required for half the atoms of a quantity of a radioactive isotope to decay (to be converted either into another isotope of the same chemical element or into another chemical element) and hence the time for the radioactivity to be reduced to half its initial value.

Halocarbons: Chemical compounds with methane-like structure, where hydrogen is substituted by chlorine and/or fluorine. The halocarbons

11 and 12 (or Freons-11 and -12 according to the nomenclature of one major manufacturer) are chemically inert like noble gases.

Heterotrophs: Organisms that take up carbon from the environment as higher organic compounds – animals, fungi, and most bacteria. The opposite: autotrophs.

Isotopes: Atoms having the same nuclear charge (the same number of protons) and thus the same chemical properties, but different mass number (different number of neutrons) and hence different nuclear and other physical properties like mobility, penetration and boiling point.

Mass spectrometer: Instrument for measuring the isotopic composition of chemical elements.

Mutation: Suddenly occurring change in a gene (inherital disposition) that is passed to all daughter cells of this particular mutated cell.

Nitrate respirers: (denitrifying bacteria) Bacteria that, in the absence of free oxygen, can utilize the oxygen contained in nitrates for 're-spiration', hence the oxidation of organic carbon compounds. Thereby nitrate (NO_3) is being reduced to either laughing gas (N_2O) or nitrogen (N_2).

Nitrifying bacteria: A group of bacteria that gain their energy through 'burning' of ammonia (NH_3) to nitrite (NO_2), and of nitrite to nitrate (NO_3), respectively.

Noble gases: Gases that are chemically inert, i.e. that do not react with other substances: helium, neon, argon, crypton, xenon, and radon, the radioactive decay product of radium.

Photolysis: The decomposition of chemical compounds through ultraviolet or visible light.

Photosynthesis: The production of higher organic compounds from simple basic substances, by means of the energy of sunlight, in particular the production of carbohydrates by green plants.

Prokaryotes: Primitive unicellular organisms that have no cell nucleus (bacteria, blue-green algae). The opposite: eukaryotes.

Radioactivity: Spontaneous decay of an unstable atomic nucleus that is thereby converted into an atomic nucleus of another isotope or another element, under emission of radiation.

Reduction (in chemistry): To take away oxygen from a chemical compound or to attach hydrogen to it. Reducing atmosphere: Atmosphere that contained more than mere traces of hydrogen and non-oxidised hydrogen compounds.

Sediments: Deposits from erosion or biological processes, in particular clay, sands and gravel. In the course of time these can solidify to form sedimentary rocks (sandstone for instance).

Stratosphere: Atmospheric layer located above the near-surface troposphere where solar energy is absorbed by photolytic processes and ozone formation, and temperature increases with increasing height. Contains (nearly) no water vapour and hence no clouds. Lower boundary in mid-latitudes about ten kilometres, in the tropics up to eighteen kilometres above sea-level, upper boundary about fifty kilometres.

Symbiosis: Close community of different kinds of species, for instance lichens (algae and fungi) or leguminous plants and Rhizobia.

Troposphere: Lowest layer of the atmosphere, where clouds exist and the weather is found. Extends up to ten kilometres at mid-latitudes, up to eighteen kilometres in the tropics.

Index

Note: Page numbers in *italics* refer to tables, figures and plates
Page numbers in **bold type** refer to large references, i.e. chapters

afforestation *see* reforestation
air, composition of, *vii–viii*, 6, *171*; as a result of soil erosion, 17; *see also* atmosphere
algae, as oxygen producers, 17–20
Amazon basin, deforestation in, 9–10
ammonia, 105, 106, 107, 109, *114*, 117
argon, *vii*, 27–8, 31, *171*
atmosphere, history of, **25–66**; banded iron formations, 50–4; dimensions of, 25–8, *75*; evidence for lack of oxygen in early times, 31–4, 51; origin of biological cycles, 58–61; origin of life, 38–41, 44–8; photolysis, 32–3; and effects of solar radiation, 34–8; temperature profile of, *37*; volcanic gas composition, 29
atomic energy *see* nuclear power
Azoto spp., 110

bacteria, *114*; primitive, 106–7, 108, 109–10; and symbiosis, 110–11
banded iron formations, 50–4
biological cycles, origins of, 58–61, *62*
biomass, **172–3**; effects of fossil fuel consumption on, 76–81
blue-green algae, 44, 45, 50, 54, 108, 109–10, 111, 112, *114*
Brazil, deforestation in, 9, 10

calcium carbonate, in oceans, 71–2
carbon-14, 72–4
carbon cycle, global atmospheric, *2*, 3, *18*
carbon dioxide, *vii*, *viii–ix*, 30, 137, 145, 148–51; and biomass fluctuations, 76–81; concentration of, 3, *4*, 5, *83*, *171*; deforestation effects on, 7–13; fossil fuel consumption effects on, 67–76; future trends, 89–96; global temperature effects on, 96–104; risks of increase in, 7–13, **67–104**; solar radiation effects on, 35; and tree-ring evidence 81–4; in volcanic emissions, 6, *29*
carbon reservoirs, 3–6, 19
chlorofluoromethanes, and effect on ozone layer, 100, 127–8
Clostridia spp., 107, 110
combustion products, *ix*

deforestation, *viii*; and carbon dioxide increase, 7–13, 69, 85, 96; and flooding, 14; and oxygen depletion, 21; and subsequent soil erosion, 13–17
denitrifying bacteria, 113
dichlorodifluoromethane (CF_2Cl_2, Freon-12), effect on ozone layer, 127–8
dust, presence of in stratosphere, 99

energy demand, global, future trends of, 147–8
erosion, and oxygen loss, 13–17, 65–6
eukaryotes, 56, 108
eutrophication, 87

fermentation processes, of early life forms, 41–3, 107
fertilisers, nitrogen, *ix*, 115–19; and increase in atmospheric nitrous oxide, 100, 127
flooding, as consequence of deforestation, 14
forests, 1–7; primary production and biomass of, *172–3*
fossil fuel consumption, and risk of increased carbon dioxide, 67–76, 138, 144
Freon-11, Freon-12, and effects on ozone layer, 127–8

gene transfer, 109
geothermal energy, 152
grasslands, primary production and biomass of, *172–3*
greenhouse effect, 96, 118, 128; *see also* temperature, global

halocarbons, and effect on ozone layer, 127–9
Himalayas, deforestation of lower ranges, 11, 14, 16
humus destruction, 7, 9
hydrogen, *vii*, 26, 29, 30, *171*
hydrosphere, early history of, 30

ionosphere, 26, 35

Klebsiella spp., 109

laughing gas *see* nitrous oxide
lichens, 110
life, origins of, 38–41, 44–8, 105–8

manure, and soil recovery, 116
methane, *vii*, 29, *170*
methanol, as liquid fuel, 157
Mexico, deforestation in, 11, 15

nitrates, 113
nitric oxide (NO), 117, 121, 125–6
nitrifying bacteria, 113, *114*
Nitrobacter spp., 113
nitrogen, *vii*, *ix–x*, 31, *171*; origins of, in atmosphere, 105–8; sink, in ocean, 19; in volcanic emissions, *29*; *see also* nitrogen cycle, individual nitrogen oxides
nitrogenase, 106, 107, 108, 109
nitrogen cycle, *ix*, *x*, 111–15; flow rates of, *115*
nitrogen dioxide (NO_2), 117
nitrogen fixation, 107, 108, 109–10, *115*
Nitrosomonas spp., 113
nitrous oxide (N_2O, laughing gas), *x*, *114*, *115*, 117–18, 119–25, *171*; and effect on ozone layer, 125–7; increase of, due to fertilisers, 100, 120, 123–4
nuclear power, 138–44, 148, 150, 152–3, 154

oceans: calcium bicarbonate in, 71–2; primary production and biomass of, *172*; as reservoir for carbon dioxide, 71, *75*, 87–8
ocean sediments: and carbon reservoir, 6; and nutrient upwelling areas, 19
oxygen, *vii*, *viii*, 21–4, *171*; and biological cycles, 61–5; formation of, in early times, 55–8; loss of, by erosion, 13–17
oxygen consumption, global, *viii*, 12
ozone layer, *x*, 25, 36, *37*, *171*; and effect of chlorofluoromethanes on, 100, 125, 127–9; formation of, 35–6, 55; and effect of nitrogen oxides on, 121, 125–7; reduction of, and increased incidence of skin cancer, 130–2; and effect of supersonic aircraft on, 127

phosphates, as fertilisers, 117
phosphorus, sink in ocean, 19
photolysis, 30, 32–3
photosynthesis, 1, 3, 7, 43–4, 46, 60, 63, 107, 108; and banded iron formations, 50–4; in oceans *see* algae
pitchblende, 32
plutonium, 138–40, 142, 143
pollution, oceanic, and marine algae destruction, *viii*, 20
primary production, *172–3*
primordial soup, 39, 41, 106
prokaryotes, 44, 108
pyrite (FeS_2), 31

reforestation, 7, 10, 13, 159–66
respiration, 1, *2*, 60, 107

savannas, 8, 9, *172–3*
skin cancer, incidence of, and effects of ozone layer, 130–2
solar energy, utilisation of, 151, 152, 153, 154–6
solar radiation: effects on atmosphere, 34–8; penetration depth, *37*
solar wind, 36
stratosphere, 25; absorption of radiation energy by, 36, *37*; presence of dust in, 99
Suess effect, 74
sulphur dioxide, 31; in volcanic emissions, *29*
supersonic aircraft, and effect on ozone layer, 127

temperature, global, and effects of increased carbon dioxide on, 96–104
temperature profile, of atmosphere, *37*
thucholithe, 47–8

transpiration, 10
tree rings, analysis of, 81–4; carbon-14 in, 74, 82
trichlorofluoromethane ($CFCl_3$, Freon-11), effects on ozone layer of, 127–8
troposphere, 25, 36, *37*

ultraviolet radiation, absorption of, by atmosphere, 35–6; and effects on life forms, 36, 38; and synthesis of macro-molecules, 111–12
upwelling, oceanic, of nutrients, 19
Urey effect, 33

vegetation, land: carbon dioxide from fossil fuel consumption in, *75*, 76; *see also* biomass
volcanoes: composition of volcanic gases, *29*; evolution of carbon dioxide, 6

water vapour: and solar radiation effects on, 35; in volcanic emissions, 29

X-rays, absorption of, by atmosphere, 35